T-Labs Series in Telecommunication Services

Series Editors

Sebastian Möller, Quality and Usability Lab, Technische Universität Berlin, Berlin, Germany

Axel Küpper, Telekom Innovation Laboratories, Technische Universität Berlin, Berlin, Germany

Alexander Raake, Audiovisual Technology Group, Technische Universität Ilmenau, Ilmenau, Germany

T0172267

More information about this series at http://www.springer.com/series/10013

Falk Ralph Schiffner

Dimension-Based Quality Analysis and Prediction for Videotelephony

 Springer

Falk Ralph Schiffner
Technische Universität Berlin
Oranienburg, Germany

ISSN 2192-2810   ISSN 2192-2829   (electronic)
T-Labs Series in Telecommunication Services
ISBN 978-3-030-56572-5   ISBN 978-3-030-56570-1   (eBook)
https://doi.org/10.1007/978-3-030-56570-1

This Springer imprint is published by the registered company Springer Nature Switzerland AG
The registered company address is: Gewerbestrasse 11, 6330 Cham, Switzerland

Something for nothing does not exist!
You always have to pay the price!
—Jim Rohn

We are all self-made,
but only the successful will admit it.
—Earl Nightingale

Declaration

Hiermit erkläre ich, Falk Ralph Schiffner, an Eides statt gegenüber der Technischen Universität Berlin, dass die vorliegende, dieser Erklärung nachstehende, Arbeit mit dem Titel

Dimension-Based Quality Analysis and Prediction for Videotelephony

selbstständig und nur unter Zuhilfenahme der im Literaturverzeichnis genannten Quellen und Hilfsmittel angefertigt wurde. Alle Stellen der Arbeit, die anderen Werken dem Wortlaut oder dem Sinn nach entnommen wurden, sind kenntlich gemacht. Ich reiche die Arbeit erstmals als Prüfungsleistung ein.

Berlin, Germany
August 2020

Acknowledgments

"The human mind ... doesn't care what we plant,
—success or failure...,
but what we plant, it will return to us."
—Earl Nightingale

My gratitude goes to my supervisor Prof. Dr.-Ing. Sebastian Möller. His constant support has been of undeniable value during all the years at the Quality and Usability Lab. As a mentor guiding my work and career, he pushed my scientific ventures to a higher level. In this context, I would like to thank him for introducing me to the work of the ITU-T Study Group 12 and the chance to participate in the standardization process. I am particularly grateful to Prof. Dr. Matthias Hirth and Prof. Dr.-Ing. Ulrich Reiter for reviewing my thesis.

My acknowledgment also goes to my colleagues, who provided a friendly and uniquely cooperative workplace. It was always a pleasure to work with you all. Special thanks go to my "seat neighbors" Gabriel(s), for the excellent time spent next to each other, and for answering the 1000 little questions. I would also like to thank Steven Schmidt and Saman Zadtootaghaj for their enriching contribution. I am also very grateful to Robert Greinacher and David März, for their remarkable support, especially with teaching. A big thank you also goes to Arne Kuhle for his help and work, especially in the quality prediction process.

Not to forget, my sincere gratitude goes to Irene Hube-Achter and Yasmin Hillebrenner for their administrative support for my research and teaching.

Finally, and most importantly, I genuinely thank my wife Christin and my children Heidelind and Thore. Standing by me in moments of doubt, they are the main drivers for my motivation.

To all the students who try to get a doctoral degree, "do not stop and keep walking," you will reach the finish line.

Berlin, Germany   Falk Ralph Schiffner
August 2020

Publications

Parts of this Dissertation have already appeared previously partly in the following publication:

[1] Schiffner, F. and Bondarenko, V. and Möller, S. "Investigation of Video Quality Dimensions for different Video Content." In: *The 20th IEEE International Symposium on Multimedia (ISM)*. electronic. IEEE, TWN-Taichung, 2018.

[2] Schiffner, F. and Möller, S. and Köster, F. *Perceptual Dimensions of Video Quality in the Context of Video-Telephony: Methodology, Experiments, Analysis*. ITU-T Study Group 12—Contribution C.95. International Telecommunication Union—Telecommunication Standardization Sector, CH-Geneva, 2017.

[3] Schiffner, F. and Möller, S. "Audio-Visuelle Qualität: Zum Einfluss des Audiokanals auf die Videoqualitäts- und Gesamtqualitätsbewertung." German. In: *42. Jahrestag für Akusitk (DAGA)*. GER-Aachen: Deutsche Gesellschaft für Akustik (DEGA e.V.), 2016.

[4] Schiffner, F. and Möller, S. "Diving Into Perceptual Space: Quality Relevant Dimensions for Video Telephony." In: *2016 8th International Conference on Quality of Multimedia Experience (QoMEX 2016)*. PO-Lisbon: IEEE—Signal Processing Society, 2016.

[5] Schiffner, F. and Möller, S. "Defining the Relevant Perceptual Quality Space for Video and Video Telephony." In: *2017 9th International Conference on Quality of Multimedia Experience (QoMEX 2017)*. GER-Erfurt: IEEE—Signal Processing Society, 2017.

[6] Schiffner, F. and Möller, S. "Direct Scaling & Quality Prediction for perceptual Video Quality Dimensions." In: *2018 10th International Conference on Quality of Multimedia Experience (QoMEX 2018)*. IT-Sardinia: IEEE—Signal Processing Society, 2018.

[7]   Schiffner, F. and Möller, S. *Investigation of Video Quality Dimensions for different Video Content*. ITU-T Study Group 12—Contribution C.292. International Telecommunication Union—Telecommunication Standardization Sector, CHGeneva, 2018.

[8]   Schiffner, F. and Möller, S. *Proposal for a Draft New Recommendation on Dimensions-based Subjective Quality Evaluation for Video Content*. ITU-T Study Group 12—Contribution C.291. International Telecommunication Union—Telecommunication Standardization Sector, CH-Geneva, 2018.

[9]   Schiffner, F. and Möller, S. *Dimension Measurement of the Perceptual Video Quality Space*. ITU-T Study Group 12—Contribution C.439. International Telecommunication Union—Telecommunication Standardization Sector, CH-Geneva, 2019.

[10] Schiffner, F. and Möller, S. "Perzeptive Audio- Qualitätsdimensionen im audiovisuellen Kontext." In: *45. Jahrestag für Akusitk (DAGA)*. another medium. GERRostock: Deutsche Gesellschaft für Akustik (DEGA e.V.), 2019.

[11] Schiffner, F. and Möller, S. Quality Modelling of Audio, Video and Overall *Quality Using Perceptual Quality Dimensions in a Video Telephony Context*. ITU-T Study Group 12—Contribution C.362. International Telecommunication Union—Telecommunication Standardization Sector, CH-Geneva, 2019.

In [2–11], all the main work was done by myself. This includes designing the experiments, processing of the test material, setting up required technical systems, conducting the experiments, analyzing the data, and writing the publication. In [1], I took part in designing the experiment, conducting of the test, partially in the data analysis and wrote mainly the publication.

Abstract

Videotelephony and video transmission, in general, makes a high amount of today's internet traffic. In this sense, assessing the quality of the transmitted video is a fundamental requirement for system and service developers to ensure a certain level of quality. Moreover, the classification and monitoring of their products are of undeniable importance, especially in a highly competitive market place. However, it is not enough to provide information about the overall quality, but also to provide an insight in the root cause of low quality. So diagnostic information is needed to point to the quality lowering impairments as essential. From this perspective, traditional quality assessment methods are limited in this sense that one only obtains overall quality scores. This work presents a multidimensional analysis of transmitted video, mainly in the domain of videotelephony. The target is to examine the underlying composition of video quality and to develop the perception space relevant to video quality. This investigation should help to provide diagnostic information about the source of suboptimal quality and to deepen the understanding of the quality rating process. For this, a series of subjective experiments were conducted. The first experiments are aiming to unveil the perceptual video quality dimensions. After the identification, a new analytic method that allows gathering dimension ratings directly from naïve test subjects is presented. The method is evaluated in further experiments afterward, in the domain of video telephony and video in general. The dimension ratings gathered with the new test method are used to determine the overall video quality. Therefore, a linear quality model was developed. Further, the interaction of the perceptual video quality space with the perceptual audio/speech quality space is investigated. A first approach to instrumentally predict the perceptual video quality dimension by using video quality indicators is presented. The instrumental modeling of the overall video quality from predicted video quality dimension ratings and the fusion with a diagnostic approach from the speech telephony domain marks the end of this work. Summarizing, the multidimensional analysis of the video transmission is a further step to deeply analyze the video quality for diagnosis and optimization of video transmission systems and services.

Zusammenfassung

Videotelefonie und Videoübertragung machen im Allgemeinen einen hohen Anteil des heutigen Internetverkehrs aus. In diesem Sinne ist die Beurteilung der Qualität des übertragenen Videos eine Grundvoraussetzung für System- und Dienstentwickler, um eine bestimmte Qualität sicherzustellen. Darüber hinaus ist die Klassifizierung und Überwachung ihrer Produkte von unbestreitbarer Bedeutung, insbesondere in einem hart umkämpften Markt. Es reicht jedoch nicht aus, nur Informationen über die Gesamtqualität zu erhalten, sondern auch einen Einblick in die Ursachen für schlechte Qualität zu gewähren. Daher werden diagnostische Informationen benötigt, um auf die qualitätsmindernden Beeinträchtigungen hinzuweisen. Aus dieser Perspektive sind traditionelle Qualitätsbewertungsmethoden in diesem Sinne begrenzt, da man nur Gesamtqualitätsbewertungen erhält. In dieser Arbeit wird eine mulitdimensionale Analyse von übertragenem Video, hauptsächlich im Bereich der Videotelefonie, vorgestellt. Ziel ist es, die zugrunde liegende Zusammensetzung der Videoqualität zu untersuchen und den qualitätsrelevanten Wahrnehmungsraum zu entwickeln. Dies soll helfen, diagnostische Informationen über die Quelle nicht-optimaler Qualität bereitzustellen und das Verständnis des Qualitätsbewertungsprozesses zu vertiefen. Hierzu wurde eine Reihe von subjektiven Experimenten durchgeführt. Die ersten Experimente zielen darauf ab, die Dimensionen der wahrnehmbaren Videoqualität zu ermitteln. Nach der Identifizierung wird eine neue Analysemethode vorgestellt, mit der Bewertungen direkt von unerfahrenen Testpersonen erfasst werden können. Die Methode wird anschließend in weiteren Experimenten im Bereich der Videotelefonie und des Videos im Allgemeinen evaluiert. Die mit der neuen Testmethode erfassten Dimensionsbewertungen werden verwendet, um die Gesamtvideoqualitöt zu bestimmen. Hierfür wurde ein lineares Qualitätsmodell entwickelt. Ferner wird die Wechselwirkung des wahrnehmbaren Videoqualitätsraums mit dem wahrnehmbaren Audio-/Sprachqualitätsraum untersucht. Ein erster Ansatz zur instrumentellen Vorhersage der Videoqualitätsdimension unter Verwendung von Videoqualitätsindikatoren wird vorgestellt. Die instrumentelle Modellierung der Gesamtvideoqualität aus vorhergesagten Bewertungen der Videoqualitätsdimensionen und die Fusion mit einem diagnostischen Ansatz aus dem Bereich der Sprachtelefonie markiert das Ende dieser Arbeit. Zusammenfassend ist die mehrdimensionale Analyse der Videoübertragung ein weiterer Schritt zur teifergehenden Analyse der Videoqualitätsbewertung zur Diagnose und Optimierung von Videoübertragungssystemen und -diensten.

Contents

Acronyms

ACR	Absolute Category Rating
AP	Antonym Pairs
AQ	Audio Quality
AQD	Audio Quality Dimensions
CI_{95}	95% Confidence Interval
COL	Coloration
CP	Component
dBov	Decibel overload, relative to the maximum point of a digital system before clipping occurs
DIAL	Diagnostic Instrumental Assessment of Listening quality
DIC	Discontinuity
DSCAL	Direct Scaling
DVD	Digital Versatile Disc
FEC	Forward Error Correction
FFT	Fast Fourier Transformation
FIR	Finite Duration Impulse Response
FL	Factor Loadings
FPS	Frames per Second
FRA	Fragmentation
Fs	Sampling Frequency
GoP	Group of Pictures
H.264	MPEG-4 Part 10, Advanced Video Coding
HD	High Definition
HP	High Performance
HVS	Human Visual System
IP	Internet Protocol
IPTV	Internet Protocol Television
ITU-R	International Telecommunications Union—Radiocommunication Sector

ITU-T	International Telecommunications Union—Telecommunication Standardization Sector
LOU	Suboptimal Loudness
LUM	Suboptimal Luminosity
MAD	Medium Absolute Deviation
MB	Macro Block
MCL	Most Comfort Level
MDS	Multi-Dimensional Scaling
MNRU	Modulated Noise Reference Unit
MOS	Mean Opinion Score
MSE	Mean Squared Error
MV	Motion Vector
NB	Narrow Band
NOI	Noisiness
PC	Paired Comparison
PCA	Principal Component Analysis
PESQ	Perceptual Evaluation of Speech Quality
PL	Packet Loss
PLC	Packet Loss Concealment
PSNR	Peak Signal-to-Noise Ratio
Q	Quality
QoE	Quality of Experience
RGB	Red Green Blue
RISV	Reference Impairment System for Video
RMSE	Root Mean Square Error
RTCP	Real-Time Control Protocol
RTP	Real-Time Transport Protocol
SD	Semantic Differential
SSIM	Structural Similarity
Stdev	Standard Deviation
STL	Software Tool Library
SWB	Super Wide Band
TCP	Transmission Control Protocol
UCL	Unclearness
UDP	User Datagram Protocol
VGA	Video Graphics Array
VoIP	Voice-over-IP
VQ	Video Quality
VQD	Video Quality Dimensions
VQI	Video Quality Indicators
WB	Wide Band
WB-PESQ	Wide Band-Perceptual Evaluation of Speech Quality
YCbCr	Luminance, Component blue Difference, Component red Difference

List of Figures

List of Tables

Chapter 1
Introduction

1.1 Motivation

Interaction through communication is one of the basic necessities of human beings. Without the ability to communicate with other humans, our psychological fibers have the potential to be seriously damaged. Even more, the social constructs human societies are made of are not possible without communication.

The development of telecommunication, like speech telephony, was driven by that basic need. Videotelephony, regarded as the next logical step, is not a new concept. The first video telephony connection was set up by AT&T and the Bell Laboratories [1]—as early as the mid-1920s. Because of the technological difficulties, it took several decades, with video telephony always there, but in a niche existence. After the three "revolutions" of electronic communication (*telephony, television* and *computers*) [2], the technology to finally establish widely used videotelephony services, is developed.

For a long time, the acceptance of enormous quality problems was a prerequisite [3]. Since broadband internet connection became mainstream, the usage of videotelephony has exploded. Even more, when service providers like *Skype* offered their services basically for free, videotelephony became embedded in daily life. The gain in processing speed and technological development, like advanced compression algorithms, made it possible to provide communication in a quality the masses now accept. Furthermore, the massive reduction in cost for the users makes the use of videotelephony, even across long distances, highly effective [3]. Technological development does not end here. The building up of new mobile communication infrastructure made it possible to video call even independent from the line held network connection. In opposite of today, in bygone days videotelephony was more like a *we can do it*-thing. Nowadays, all forms of video content transmission (Internet video, IP Video-on-Demand, video-streamed gaming, and videotelephony) makes up the most significant amount of traffic online, and it is expected to rise. In total,

© The Author(s), under exclusive license to Springer Nature Switzerland AG 2021
F. Schiffner, *Dimension-Based Quality Analysis and Prediction
for Videotelephony*, T-Labs Series in Telecommunication Services,
https://doi.org/10.1007/978-3-030-56570-1_1

the video traffic will account for 82% of the traffic by 2022 [4]. Even though video telephony is only a part of the total video transmission, it is playing an increasingly important role, especially in the business environment [5]. In [5], the authors summarizes "... *video conferencing is here to stay as a key part of video-first workplace cultures, connecting distributed teams and improving employees' day-to-day workflows. Video has been a business advantage up to this point, but soon it will become an essential component of business success.*"

Because of this growing importance, in this work by video is meant, mainly video transmission in videotelephony. With this in mind, service providers and technology developers are becoming more and more concerned about the quality of their products. In order to provide the best possible quality to the user and to ensure that they stay satisfied, the perceived quality needs to be measured. To meet that need, subjective tests were developed to judge the communication systems and to understand what the user considers "good" quality. With the help of these subjective tests, valuable results were obtained, but at a high cost in time and finances. To tackle that, quality models were derived from the findings to predict the expected quality from instrumental measures.

The large amount of data currently being transmitted will continue to demand the improvement of services and technology. Even if one has access to details about quality assessment and user expectations, there are still blank spots left on the map.

This work is an attempt to fill out the map a bit more. Understanding how *video quality judgment* is formed and what it is compound of, motivated this work.

1.2 Objectives and Scope

In video quality research, passive subjective experiments in a laboratory setting are used to understand the Quality of Experience (QoE). The majority of the studies, in this domain, resulting in only a set of Mean Opinion Score (MOS). The MOS represents the average overall quality rating of the condition under test. It is common knowledge that these studies are time and financially costly. In fact, service providers demanding instrumental models to predict the quality of their products. This demand leads to several types of approaches (see Sect. 2.7). Nevertheless, these models have a main drawback; they only give an overall quality score in the end, and no further information on the cause of the potential quality loss can be derived from it. Because different types of impairments have different perceptive effects resulting in different quality ratings, it makes sense to study the composition of the overall quality. This leads to a more detailed investigation of possible root causes for a reduced quality rating. So, splitting into a multidimensional perceptual space can give a user-centered insight into the cause of the degradation. From this, in turn, the technical causes and measures for improvement can be derived. Thus the main research question for this work is

> *What are the underlying perceptual quality dimensions for video? How is the overall video quality composed of these dimensions?*

This work presents a systematic approach to answer this question. The subsequent steps are first, identifying the quality dimensions, and second, a method, already existing in speech telephony for quantifying the identified perceptual dimension, is adopted for video assessment. Further, this work addresses the modeling of the video quality in the domain of videotelephony. From that derives the second research question:

> *Is it possible to model the overall video quality from the underlying video quality dimensions?*

Since the main demand is the instrumental estimation of quality ratings, the next research question to tackle is

> *Is it possible to estimate the ratings of each video quality dimension separately and model the overall video quality from the estimated dimension ratings?*

In video telephony, not only the video channel is of importance, but also the audio channel. This leads to the last investigated research question.

> *Is it possible to assess the audio and video quality dimensions at the same time, and how do they interact? Further, is it possible to combine an instrumental-dimension-based audio quality estimator with a dimension-based video quality estimator to predict the overall quality?*

This work provides for the first time, a method and an instrumental-dimension-based video quality model, which allows analyzing video quality further. It also allows the possibility to derive more in-depth diagnostic information and gain a better understanding of the video quality judgment process.

1.2.1 Structure of the Thesis

This work is structured as follows:

Chapter 2 provides the fundamental knowledge needed for the presented work. The introduction explains the fundamentals of audiovisual perception and signal transmission. The central part of this chapter is the explanation of what quality is and how it is constructed. Further, it explains the necessary assessment methods used for this work.

In Chap. 3, a summary of the experimental setup is given. The methods explained in the previous chapter are used and modified here for the use of the investigation of video quality. The results of the conducted experiments are given together with the analysis. The chapter closes with unveiling the underlying perceptual video quality dimensions.

While the perceptual video quality dimensions are unveiled at this point, a method already used in speech telephony was adapted for the use in video quality assessment. The method itself and a series of three experiments are described to validate the approach in Chap. 4. This work is also placed in the domain of videotelephony; a view on the applicability of the method in a broader video domain as well as the perceptual video quality dimensions is provided.

In the following, Chap. 5 presents a dimension-based linear video quality model. This chapter continues with the validation of the model by applying it to several data sets and closes with the conclusion.

Based on the finding from the previous chapters, an instrumental-dimension-based video quality estimation is developed using video quality indicators. The separate estimation of the five video quality dimensions is presented in Chap. 6. Using the video quality model presented in Chap. 5, the overall video quality is estimated by using the separately estimated dimension ratings.

Chapter 7 investigates the interaction between the perceptual video quality dimensions and the perceptual audio quality dimensions. The presented study helps with the results to predict the audio, video, and overall quality. This chapter closes with a conclusion on the usage of the method presented in Chap. 4 in an audiovisual setting.

Chapter 8 presents a fully instrumental estimation of the audio-visual quality. Therefore, a tool (DIAL) for instrumentally estimating the audio quality dimension is used for overall quality modeling. The results from DIAL, and the results from the previous chapters are combined to calculate the overall quality.

Chapter 9 provides the conclusion and future work. The second to last section summarizes the research. This work closes with an outlook in Sect. 9.3. Here, a glance into possible directions this work could be developed is provided.

References

1. Dunlap Jr., O.E.: The Outlook For Television. Harper & Brothers Publishers, USA, New York (1932)
2. Schaphorst, R.: Videoconferencing and Videotelephony-Technology and Standards. Artech House, USA, Norwood, MA (1996)
3. Firestone, S., Ramalingam, T., Fry, S.: Voice and Video Conferencing Fundamentals. Cisco Press, USA, Indianapolis, MA (2007)
4. Cisco: Cisco Visual Networking Index: Forecast and Trends, 2017–2022. https://www.cisco.com/c/en/us/solutions/collateral/service-provider/visualnetworking-index-vni/white-paper-c11-741490.html. Last visit 09/2019 (2019)
5. Lifesize Inc.: 2019 Impact of Video Conferencing Report. https://www.lifesize.com/en/ldp/the-future-of-video-communication-and-meetingproductivity. Last visit 11/2019 (2019)

Chapter 2
Related Work and Theoretical Background

2.1 Audio-Visual Perception

This section presents a brief description of the perception of audio and video. In general, the perception of a stimulus can be divided roughly into four major steps. This holds not only for audio and video stimuli.

1. Stimulus arriving at the receptor,
2. Transduction into bio-electrical signals,
3. Propagation of the signals into cognitive areas, and
4. Processing and interpretation in higher cognitive areas.

2.1.1 Perception of Video

The task of the Human Visual System (HVS) is to process all information coming upon the eye and renders it into something recognizable for the higher areas of the human brain [1]. The human eye is the first stage of visual processing. It can be regarded as a bandpass filter, where only light with a wavelength between 400 and 700nm passes. It is the home for two kinds of photoreceptors (rods and cones). The rods are generally more sensitive in low-light conditions, whereas cones are sensitive to well-lighted conditions and are responsible for color vision. When hit by light photons, these receptors emit action potentials, which are fed through a complex system of nerve fibers and nuclei and finally sent to the visual cortex. The visual cortex is organized, as common is neural processing, into hierarchical areas (V1–V5). Here the final interpretation of the viewed is built. The area V5 is responsible for processing motion information [1]. The estimation of motion is of great importance for humans because here velocity, distances, etc., are used to orientate themselves within their environment. The same area is triggered when watching a video, since

© The Author(s), under exclusive license to Springer Nature Switzerland AG 2021
F. Schiffner, *Dimension-Based Quality Analysis and Prediction for Videotelephony*, T-Labs Series in Telecommunication Services, https://doi.org/10.1007/978-3-030-56570-1_2

our visual cortex is hard-wired to concentrate on moving objects. This work deals mainly with video and the technical equivalent to the motion in nature would be a flow of separate images, as it is determined in the frame rate. Modern video codecs often try to implement some degree functionality of the HVS to make video coding more efficient. For a more detailed description of HVS and its functions see [2, 3].

2.1.2 Perception of Audio

The hearing system of the human, similar to the visual system, is a complex sensory system. The peripheral auditory system is built of three main parts; the outer, middle, and inner ear [4]. The function of the outer ear is mainly to impress directional characteristic to the sound event. Also, some frequency ranges increase due to resonances at the auricle and the ear channel. The middle ear acts as an impedance converter making the transition from sound propagation in the air to sound propagation in liquid possible. Without this mechanism, a loss would be imposed through direct reflection. The leverage of the three small bones, together with the surface transformation of the relatively large eardrum to the smaller oval window, will lead to a signal gain [5]. After entering the cochlea in the inner ear, the sound wave propagates along the basilar membrane. Here, the Corti organ, with its inner and outer hair cells, is placed. In these cells, the transduction of the physical sound wave into action potentials takes place. The place on the basilar membrane where the action potentials are generated, together with their pattern, will later be used to interpret the signal. The now neuro-electric signals are sent via the auditory nerves to the primary auditory cortex. It is still not entirely clear how the signals propagate further in the brain, but studies suggest that there is a secondary auditory cortex followed by the associative auditory cortex.

For more details on audio perception and hearing in general, see [3–5].

2.2 Signal Transmission

The invention of the *moving picture* is one of the most influential upbringings of the past century. All its incarnations, like cinema, television, video-streaming, videotelephony, had a significant impact on society. The transition from analog video to digital video processing has been completed, and analog video can only be found for nostalgia or artistic reasons. The advancing technology, especially powerful compression algorithms, made it possible to send and receive video data everywhere. With this help, the bandwidth and storage needed could be reduced when regarding a single video signal on its own. The "drawback" of such powerful systems and the possibility to use them 24/7 cause a "problem", which is that today's internet traffic consist heavily on video transmission. The latest Cisco forecast estimates video traffic to be approximately 82% of the total internet traffic by 2022 [6]. Since this is also a significant business factor, companies prioritize designing new digital video systems, and service providers are more concerned about guaranteeing a certain level of quality.

In this section, a brief introduction about data transmission is provided. A particular focus is put on the video and audio–video data transmission and its potential errors.

2.2.1 Video and Audio–Video Data Transmission

The data transmission via the Internet relies mainly on two protocols [7], namely, Internet Protocol (IP) and Transmission Control Protocol (TCP). There, it establishes no direct line connection from the sender to the receiver. In IP, it is specified that all information going to be sent via the network is divided into data packets and marked with an "electronic envelope," so to speak. With the help of the envelope, all necessary information about the address of the packet and its order in the overall information is sent. Every transmission knot (router) helps in organizing to find the correct path through the network. The reliable TCP is not suitable for real-time video transmission. In the event of a packet loss, the missing packets are retransmitted, and the resulting delay is unacceptable. For this reason, the unreliable User Datagram Protocol (UDP) protocol is used here. There is no guarantee that the data streams will reach the receiver. Therefore, the Real-Time Transport Protocol (RTP) and Real-Time Control Protocol (RTCP) are used here. RTP is an internet standard protocol designed for the end-to-end transmission of real-time applications. The RTCP is an RTP accompanying protocol that gives session participants feedback of an RTP session (e.g., about packet loss).

After the encoding of audio and video streams, the signals will be prepared for transmission. Therefore, the audio and video streams are packetized and get timestamps in the synchronization layer. The synchronization layer packets are handed over to RTP, UDP, and IP. The resulting IP-packets are now transmitted via the Internet. The transmission of the packetized streams can occur separately or together (multiplexed) [8]. The IP-packets are numbered and the sequence number count of each RTP-packet. This helps to reorganize the order of the packets at the receiver or identify missing packets and eventually send a new sending request (in TCP). Also important to take into consideration are the timestamps for the synchronization of audio and video for a *lip-sync* playback. The Maximal Transfer Unit of the network defines the size of the packets and can vary. This is done to limit the negative influence of packets that are missing or corrupted. Once the packets have reached their destination, the de-packetization starts. The audio and video stream is then cached in a jitter buffer. The task of that buffer is to order the packets according to their sequence number. Since the playout time and thus, the overall delay is dependent on the buffer size, the jitter buffer can often be adjusted dynamically in order to reduce the overall delay. Nevertheless, a too-small buffer has the drawback that it is prone to network jitter. If the order is finally correct, the stream will be sent to the decoder and further sent e.g. to the screen and speaker. The final presentation of the audio, video, or audio-visual material depends now on the settings of the playback system (e.g. color schema of the screen, size of playback window, frequency range of the speaker). As a side note, what should be taken into consideration is that the sending and receiving sides have to agree about a few general conditions, like which algorithms and protocols are used to enable interoperability [9]. One solution to interoperability could be the transcoding of the stream. However, this could lead to a significant drop in quality. To circumvent that, it is beneficial to agree on common codecs and protocols beforehand.

2.2.2   Transmission Errors

The digital audio-visual signal is handed over to a packet-switched network. This can be a wire-based or wireless network. Therefore, it is unlikely that the audio and video streams are sent via the same path to the receiver [8], and even single IP packets or groups of packets are sent via a different path. To illustrate that process, Fig. 2.1 is given. During this process, it frequently happens that packets become missing, corrupted, or get delayed. Late or irregular arriving packets can lead to errors (e.g. interruptions in the flow of stream). If a packet arrives too late to be recognized or is lost on the pathway, this type of error is called Packet Loss (PL). PL can occur due to the packet routine and queuing algorithms in the routers and switches. The negative impact of packet loss depends on several aspects. Among other things, the loss rate, the loss distribution, and the packet size matter. All the

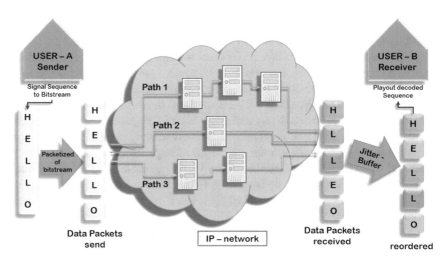

Fig. 2.1 Schematic depiction of packet-oriented data transmission. The signal sequence is encoded, and the resulting bitstream is packetized and sent to the network. The received packets are reordered and stored in the jitter buffer, depacketized and decoded for the playout

major types of transmission errors (delayed, corrupted, or lost packets) have more or less the same consequence: The information is not available for decoding. Such losses can affect e.g. a whole video frame or only parts of it. The emerging artifacts can affect not only the current image but also the other frames that are dependent on the information. Since a video is arranged in a so-called Group of Pictures (GoP) [10], the loss of an I-frame is more damaging than from a P-frame. This is due to the fact that the information from an I-frame is used to decode P- and B-frames of the whole GoP. In this manner, the loss and the artifacts propagate to the next frames. The loss of a P-frame has more impact on B-frames because its information is used to decode other P- and B-frames. The lowest impact has a lost B-frame because it is not used to decode other frames. It impacts only itself.

In general, it can be said that the loss effects will be propagated until the following information for decoding is available [11]. The visibility of such artifacts varies significantly. They not only depend on the packet/s lost but also on the implementation of the decoder.

2.2.3   Recovery Mechanisms

Before the playout, some recovery techniques are commonly implemented to compensate or at least reduce the effects of corrupted or lost packets. To reduce the need for compensation efforts on the receiver side, Forward Error Correction (FEC) and packet retransmission is used on the sender side. The idea behind FEC is to include some redundant information (like parity bits) into the signal [12]. On the receiver side, Packet Loss Concealment (PLC) algorithms are employed. Their performance depends on the implementation of the decoder. The PLC algorithms uses correlation of neighboring Macro Blocks MBs of the same, previous or following frames. These methods can be grouped into spatial and temporal concealment algorithms (see e.g. [11, 13, 14]).

2.2.3.1   Spatial PLC

These algorithms use the spatial correlation of neighboring MBs of a video frame, intending to replace every missing pixel of the impaired part of the frame with pixels from the unimpaired surrounding. There are several methods to do so e.g.:

- MB copy: The most basic technique is to calculate the corrupted MB with interpolation from the next MB. This leads to mirror-like artifacts.
- MB mean-interpolation: The mean of the Macro Block is calculated from the correct decoded neighboring pixel. This leads to block-artifacts since all pixels of the MB is appointed to the same value.

- Directional interpolation: This method can be costly since it requires an edge detection. Here the algorithm searches for edges and their directions. The MB is then interpolated along the found edge direction.
- Maximally smooth interpolation: The algorithm uses the frequency domain to interpolate the reduced number of DCT-coefficients with the goal to make the transition of the reconstructed pixel to the next one as smooth as possible.

2.2.3.2   Temporal PLC

Here, the algorithms replace the impaired or missing pixel with a pixel from the previous or following frame. This can be done by replacing the pixel directly by copying from a previous or following frame or by replacing it by a motion-compensated one [12]. For the latter, an estimation of the Motion Vector (MV) is necessary. The performance of the method depends mainly on the goodness of that estimation. Nevertheless, a motion-compensated substitution of the impaired pixel is most desirable for video; since there is lots of motion, and simple substitution of the pixel would lead to more visible artifacts. Similar to the spatial PLC, there are several different methods for temporal PLC:

- Zero MV: This is the most straightforward approach, where the pixels are replaced with an unimpaired pixel from the previous or following frame in the same position. Here, no motion compensation takes place, or put it another way; the Motion Vector (MV) is set to zero.
- Neighboring MV: Here, the surrounding of the impaired MB is investigated for Motion Vectors. The found MVs are used to replace the lost one.
- Previous and Following MV: A method that yields excellent results is looking for MV in the previous and next frame on the same position, where the impaired MB is placed. Unfortunately, these frames are not always available.

Depending on the properties of the lost frame or Macro Block, one or another approach leads to better loss concealment. Also, a combination of all the methods mentioned above is possible.

2.3   Quality of Transmitted Video

2.3.1   Quality of Transmitted Video

The previous section gave a brief look into modern data transmission. The transmission of video and speech for video telephony contains a high load of information. Thus, for a transmission system like this, it is crucial to provide all information. In addition to classical speech telephony, it is not sufficient to have an intelligible and comprehensible speech channel, but also the appearance of the video channel also plays a huge role. Even if the standard is quite high in modern videotelephony systems, in the process of transmitting, degradations may be introduced to the video. When one tries to quantify the perception of the transmission system by the user, the perceived quality of the system is needed.

This section gives a general introduction about what quality is and how a quality judgment is formed. Furthermore, this chapter goes into more detail about video quality, audio-visual quality, and also describes some of the methods that are used for quality assessment.

2.3.2 Definition of Perceived Quality

"Today, the phone connection was terrible." "Looking at this, one can firmly state that this is of high quality." "The quality of his supervision was excellent." These or similar statements can be found all over in daily life. It seems that *quality* is a factor of enormous influence. Since a vast amount of today's technology is aimed to be used by people, "quality" is getting more and more in focus. Thus, quality plays a tremendous role in designing products and services [15], emerging from technological advancement.

If one wants to approach the concept of quality in the first step, the definition of quality from Jekosch [16] is helpful. There it says:

> *Quality is the result of a comparison between the perceived composition of an entity and the desired composition.*

This "entity" does not necessarily have to be a physical object. Likewise, a service can become this object if one can assign features to it. These features must be quality-relevant and measurable characteristics of the object. A statement about the quality is obtained by assessing the object's characteristics. Here, the expectation and the perception situation play a decisive role.

2.3.3 Subjective Quality Judgment

As the title suggests, quality can only be measured with the help of perceiving and judging persons and is a highly complex process. Quality can be seen as a result that always is dependent on the perception and assessment situation [15]. Since quality judgment is purely subjective, quality cannot be objective. It always needs, directly or indirectly, the individual.

As an example: if the object is a visual signal, as in the case of the video in videotelephony, the perceived features (color, brightness, etc.) will be detected. This allows different signals to be distinguished from each other. The composition of these features will define what the stimuli looks like ("is the swan black or white"). In order to obtain statements about the quality, a sample (e.g. a video sequence) is given to a rating instance for judgment. In the specific case of the quality of video signals, primarily the human evaluates. In this process, a variety of factors play a role. The whole process can be deconstructed in several steps (comp. Raake [17] and Jekosch [16]).

In the following, the rating process of a visual signal (e.g. video channel of videotelephony) is described, but the general process is valid for quality ratings of all kinds.

1. Optical event: Is a physical process and is determined through its physical properties (e.g. wavelength). An optical event is objective and exists independently from subjective perception. In this example, it would be the transmitted video on a screen.

2. Perception: An optical event is received by the human visual system (eyes, nerves, visual cortex, etc.). At this moment, the optical event becomes a perceptual event ("visual event"). It contains an amount of possible information (e.g. brightness, frequency content) but also information from the sending side (e.g. place of recording, atmosphere, mood of the sender, etc.). In addition, all information, emotional state, situation, experience, etc., of the receiver plays a role in the perception. For example, an identical signal is perceived differently depending on the constitution.[1] All of these factors continually adapt to the perception of what has been seen.

3. Consideration: The receiver considers all factors of what has been seen. In the process of forming the quality judgment, only the quality-relevant features are considered. This process is subconscious.

4. Comparison: The perceived features are now internally compared with the expected constitution of the features. There can be different values for different features (e.g. the brightness was as expected, but the color was dull). The distance between the values forms the basis for the subsequent step.

5. Judgment: The receiver now takes the distances from all features into account. The distances are weighted differently depending on their size and importance, and finally, the overall rating is formed. In the example (transmitted video), the color was not as expected, and therefore the distance for that feature is large. This leads to a loss of overall quality. The receiver can describe the quality (e.g. "The video in that call had a dull color (description of features), but the overall quality was fair (description of quality)").

It should be noted that the expectations can be exceeded. This would have a positive effect on the rating (e.g. expecting DVD video and watching a Blu-ray for the first time). That said, the quality judgment of "something" is not static but constantly changing. Figure 2.2 is given to illustrate the abovementioned process. It should be

---

[1] von Thun [18] has assigned to the sender and the receiver "four levels" of information (factual level, self-revealing, relationship level, appeal level). The type of interpersonal interaction depends on how the "four levels" interact.

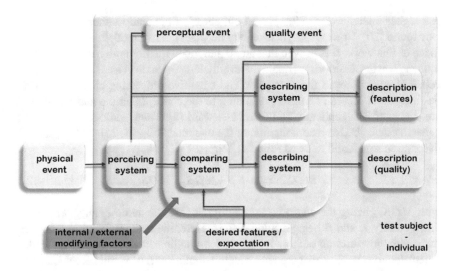

Fig. 2.2 Schematic representation of a person forming a quality rating (based [17, 19])

emphasized once again that the internal reference (desired features) of a person is particularly crucial in assessing the quality. It summarizes all aspects of the individual expectation, appropriate requirements, or even the social demands [15].

As one can see in Fig. 2.2, the quality event happens inside the individual. When working with quality judgments, one is dependent on the description of the quality event. Because the modifying factors can only be partially controlled, and the perception process is subject to individual fluctuations, quality judgments always have a degree of uncertainty. To tackle that problem, efforts are made to control as many variables as possible to reduce the room for errors.

2.3.4 Audio-Visual Integration

When regarding the different senses and how they interact, stimulus patterns from different receptors are typically aligned and matched in higher cognitive areas to form a depiction of the environment. An example would be walking a graveled path. The eyes would see the little stones and rocks, whereas the ear would hear the crunching and grinding sound from the steps. Also, the input from the feet would be a loose, uneven feeling. It is known that human senses can influence each other.

A vast number of quality metrics developed use some form of knowledge about the perception. It is also common knowledge that visual stimuli can influence the audio perception [3] and vice versa. Two factors mainly influence the audio-visual quality. The first is the synchronization between the modalities, and the second is the audio and video quality itself. It is interesting to notice that the literature reports there is a higher tolerance for video played out ahead of the audio rather than the other way round [20]. However, asynchronicity between the two modalities is outside the focus of this work. It is proven that audio and video quality both have a significant influence on the audio-visual quality. Nevertheless, for audio-visual stimuli, it is still veiled at what point in the perception and cognitive processing chain the modalities integrated. Researchers are gravitating to the concept that the audio and video are evaluated independently and fused afterward [8]. Thus, audio-visual quality is generally described as a combination of the two modalities [8, 21]. The most stable description was found to be as follows [22]:

$$MOS_{av} = \alpha + \beta \cdot MOS_a \cdot MOS_v \qquad (2.1)$$

with MOS_{av} being the audio-visual quality, MOS_a the audio quality, MOS_v the video quality, α and β being scaling factors. Here, the authors found in an experiment with video stimuli ranging from broadcast audio and video quality to standard videotelephony quality, that the video quality dominates the overall perceived audio-visual quality in a nonconversational setting. In addition, there are several studies about the cross-modal interaction of the two modalities. In a broad investigation about the relationship between audio and video in terms of finding a dominant factor, no definite pattern was found [23]. There the authors stated that audio quality and video quality are equally important in the overall audio-visual quality.

In the case of videotelephony, the influence of the audio quality level on the perceived video quality is uncertain [9], because there are contradicting study results. The influence of the video quality level on the perceived audio quality is found in three of the five studies. Summarizing, it is hypothesized that the content of the test material or the assessment method has made the difference. It is also hypothesized, if the MOS of one modality is significantly different from the other, the quality-wise better modality will correlate more with the resulting audio-visual quality [8]. This should be kept in mind when designing an experiment or analyzing the resulting data. Furthermore, it was shown that when users have to pay attention to degraded stimuli, they became more fatigued compared to high-quality stimuli [24]. When participants become mentally exhausted, they perform worse (e.g. test tasks, failed communication). This could lead to avoidance of a low-quality service or at least a less frequent usage when the quality is low and is obviously in contrast with the interests of technology and service providers.

2.4 Subjective Testing

This section presents an overview of the influencing factors of subjective tests. Many factors can influence the rating and the way test participants react in a test situation.

When determining the quality, as in the case of video transmission, a test subject leaves a description of the quality after a presented stimulus. This bears several problems. The quality perception is scaled by the subject to depict his description on a scale [17]. The transfer of the assessment on a scale provided by the test supervisor can sometimes lead to distortions. A commonly used scale is a discrete five-step Absolute Category Rating (ACR) scale, as described in the ITU-T Rec. P.800 [25]. This scale is used in many studies; however, it has some known disadvantages. The test participant may be inclined to avoid the ends of the scale since he suspects that better or worse samples could come in the test. This leads to a compression of the judgments. On the other hand, if the test participant used the ends of the scale and a poorer or better stimulus shows up, the test participant is not able to rate it properly anymore. This leads to a saturation of judgments. Another aspect is that the labels of commonly used scales should be equidistant. Nevertheless, it has been proved that this is not the case [15]. Additionally, the interpretation of the scale labels is different from person to person. This occurs within a language and also between various languages. To mitigate these effects, different scale types were developed; e.g. to get rid of the not equidistant scale labels, ruler like marks are used to imply that they are equidistant. In addition, so-called *overflow areas* are added to tackle the difficulties at the ends of the scale. An example of this kind of scales can be seen in Fig. 3.4 in Sect. 3.2.2.

When evaluating a sequence of stimuli,[2] the stimuli are not rated individually, but in relation to each other. This is why the same stimulus is rated differently in differently balanced experimental setups. Therefore, it is vital to cover a certain range of quality if the results are not only relative but also should have absolute significance. Additionally, sequence effects can occur in the assessment when the presentation order of the stimuli is the same for all test participants. This should be avoided, and a randomized presentation order should be utilized.

One can divide subjective tests into two broad categories. The first one is field testing, where the test participant undertakes tasks in a "close as possible to the real world" environment. In this case, unforeseen effects may influence results since it is not controlled. When the participant typically knows that this is a test situation, they may behave differently. The second one is laboratory testing. Here the test supervisor can control most parameters, and therefore reduce the influencing factors. The drawback of this is that the test participant is sitting in a lab room. This artificial situation and also the controlling instance (the supervisor) could lead to a different behavior of the subject. Both categories have their advantages and disadvantages, and it is up to the study supervisor to choose the right one. Nevertheless, when dealing with subjective testing, the bias of the test person plays an important role. To reduce the individual bias and to bring the test participant on the same level, various points should be addressed.

Before a test can start, participants need to understand what the test is about and what should be expected. Written instructions are therefore used. The advantage of written instructions is that every participant has identical knowledge about the test upfront. This does not guarantee that the instructions are understood correctly. If

---

[2]Here, with stimuli is meant, the presentation of the different test materials and/or degradations, but also different test sessions or test tasks (e.g. passive rating tasks or active task like conversation experiments).

the supervisor only gives the instructions, there could be interpersonal effects. A different wording could influence the behavior of the participant. However, there can also be the possibility to ask questions. In addition, to prepare a test person for the test session, training should be applied. Within the training, a certain number of different stimuli should be presented before the actual test run. With this, the participant knows the "quality range" expected in the test and can better exploit the rating scale. This also means, however, also that the subject is influenced in advance.

All factors influence the rating and the behavior of the test participant differently, and therefore a certain number of participants is required to take part in a study. With this measure, the different individual influences are averaged out to a certain degree. In the ITU-T Rec. P.800 [25] the number of participants should be 12 or more. The ITU-T Rec. P.910 [26] recommends for a video quality study to have at least 15 participants, and states that there is no additional value if the number of participants is beyond 40. The actual number of participants for a specific experiment should always be dependent on the required validity.

In summary, there is no gold standard, but a number of general rules should be followed and adapted if needed to obtain meaningful results.

Several bodies help with guidance in conducting subjective experiments, especially in the domain of telecommunication and media systems, like e.g. the International Telecommunications Union-Telecommunication Standardization Sector (ITU-T), or International Telecommunications Union-Radiocommunication Sector (ITU-R). The ITU-T Rec. P.800 [25] is meant to give guidance for audio quality test, the ITU-T Rec. P.900 [22] for video quality test, and ITU-T Rec. P.911 [22] for audio-visual tests. Based on these recommendations, all studies in this work have been structured.

2.5   Dimension-Based Approach to Video Quality

2.5.1   Perceptual Features, Dimensions and Overall Quality

As described in Sect. 2.3, a perceived physical event becomes a perceptual event. The perceptual event is generally multidimensional and consists of multiple perceptual features. When looking at the definition of "feature" from [16], it said, *a feature is a recognizable and nameable characteristic of an entity*. In the context of video in videotelephony, the entity represents the visual event, and the features could be named like e.g. brightness and color. The entity can, therefore, be deconstructed by a composition of the features. The perceptual event can now put into a perceptual multi-feature space. If the chosen configuration of the coordinate system is Cartesian, that is, when all axes are orthogonal, the feature axes become dimensions. When each perceptual feature can be linked directly to one of these orthogonal axes, these features can be regarded as perceptual dimensions. These perceptual dimensions are orthogonal, as well. The dimensionality is dependent on the number M of the perceptual features from the beginning on. Looking again at Fig. 2.2, the perceived manifestation of the features fp, and the desired manifestation of the features fd are both placed in the M-dimensional perceptual space. In this line, the overall quality can be modeled with a mapping function $func_m$.

$$Q = func_m(fp, fd) \tag{2.2}$$

The underlying perceptual dimensions define the constructed perceptual quality space. Thus, the mapping function $func_m$ defines the relationship between the underlying perceptual dimensions and the overall quality.

There are two types of dimensions that can occur and which should be taken into account (comp. [17]). First, the dimension shows a "the more, the better" behavior regarding the quality. This is often referred to as *vector model*, where the quality vector point at an optimum value. Second, the dimension has an optimum value on the axis, which is associated with the highest quality. It is often called *ideal point model*. In the domain of video quality, an example for the first would be "continuity" the more—the better. For the second, an example would be "brightness". Since there is an ideal brightness, a too low value means too dark and a too high value too bright.

To position the manifestation of one perceived feature into the multidimensional perceptual space, the concept of vectors is used. As stated, the quality judgment is a result of a comparison. Now the perceived feature fp is depicted as a vector $\overrightarrow{fp}$. At the same time, the desired manifestation of the feature $\overrightarrow{fd}$ is also placed in that room. The distance between the two, together with a weighting, forms the rating for that feature. This process is done simultaneous for all quality-relevant features and is finally added as a linear combination to the overall quality rating Q. Equation 2.3 is an attempt to explain this process more analytically:

$$Q = \alpha + \{\beta \cdot (\Delta(fp1, fd1)) + \gamma \cdot (\Delta(fp2, fd2)) + \cdots + \delta \cdot (\Delta(fpM, fdM))\} \tag{2.3}$$

In this equation, α stands for the overall bias of the rating instance like the mood or distraction of the participant. The Δ represents the difference of the perceived feature $\overrightarrow{fp}$ from the desired feature $\overrightarrow{fd}$. The other scaling factors (β, γ, δ) represent the individual importance the specific features have for that participant. M represents the number of features taken into account.

2.5.2 Dimension-Based Quality Models

Considering the quality judgment process and the creation of a multidimensional quality space, the approach follows that transmitted video quality can be predicted based on the relevant quality dimensions. To tackle that a function $func_{cs}$ that reflects the *comparison system* (comp. Fig. 2.2) must be determined. Taking into account that the comparison and the describing system are potentially in danger of bias and that they cannot be measured directly, action must be taken to reduce these biases to a minimum, as partly described in Sect. 2.4.

With that in mind, the function $func_{cs}$ can be determined in principle based on subjective ratings. Nevertheless, it cannot be guaranteed that the result is now bias free, and as a result, the $func_{cs}$ could be biased as well. When the features are orthogonal and all of them, as defined before, are relevant for the overall quality, then they can be regarded as quality dimensions. A realization of a function $func_{cs}$ as a dimension-based model will be presented in Sect. 5.1.

2.5.3 Quality Dimensions and Overall Quality—Related Work

The approach that an overall quality judgment is constructed via underlying perceptual dimensions was already used for quality modeling of transmitted speech in telephony. Besides a better understanding of how the overall quality judgment is constituted, the approach of *quality-relevant perceptual dimensions* may help to diagnose the source of suboptimal quality (e.g. jerkiness, noisiness, etc.). The quality-relevant perceptual space for speech telephony has already been mapped and dimension-based quality models were developed (e.g. [19, 27]). Here, four quality-relevant perceptual dimensions are unveiled, namely *Coloration, Discontinuity, Noisiness*, and *Loudness*. Coloration refers to degradations that impair the natural appearance of the speech signal (e.g. metallic, muffled, or nasal sound). Discontinuity is triggered by interruptions in the flow of the speech signal through, e.g. *cut outs*. Noisiness, as the names suggest, refers to noise added to the signal. This could be from the sending side or the channel. Lastly, loudness refers to either a too quiet or too loud speech signal. These dimensions were also used for the dimension-based speech quality estimation model called DIAL [27]. The proposed speech quality dimensions are also used in this work when addressing the audio part of the dimension-based audio-visual quality investigation (see Chap. 7).

In work from Köster [28], a multidimensional analysis for telephone conversations was conducted. Here, the author divides the conversation situation into three phases. In the investigation of the underlying quality dimension, the four speech quality dimensions were used in the listening phase of the conversation [28]. Further, more underlying dimensions were found for the speaking and interacting phase.

This approach was recently used in the domain of high-quality video [29]. The authors used a Paired Comparison (PC) with a subsequent Multi-Dimensional Scaling (MDS) (details on the methods are given in Sect. 2.6.2). Here, they investigated the *sharpness* and the *noisiness* of high-quality material, as can be found in professional motion picture production, and the influence on the quality rating.

In the domain of IPTV [30], the author constructed from subjective ratings a perceptual video quality space. In this work, only the video channel was of interest, and the audio channel was muted. The used method was a Semantic Differential

(SD) with as Principle Component Analysis (PCA) (details on the methods are given in Sect. 2.6.1). It unveiled three perceptual dimensions, namely, *Fragmentation, Movement Disturbance, and Frequency Content*. That work is especially interesting because it raises the question if the three dimensions can also be found and used in videotelephony.

Taking the work in the domain of speech telephony and IPTV into account leads to the question if this approach can be used in a videotelephony domain.

2.6 Experimental Paradigms

In this section, the two different experimental paradigms are presented. Both methods reduce a high-dimensional space into a low-dimensional space. For this, they follow different approaches with particular advantages and disadvantages. Section 2.6.1 describes the method of Semantic Differential (SD) with a subsequent Principle Component Analysis (PCA) using ratings for attributes. The method of Paired Comparison (PC) with a Multi-Dimensional Scaling (MDS) is introduced in Sect. 2.6.2, where dissimilarity ratings are obtained in a comparison experiment.

The two methods have one thing in common: they both are capable of transforming experimental data into a low-dimensional representation. In addition, the advantages of one method can compensate for the disadvantages of the other method to a certain degree and vice versa. Thus, the two paradigms when combined can give a solid statement about the actual nature of the underlying perceptual dimensions. This helps to verify the validity of each outcome.

2.6.1 Semantic Differential and PCA

In a Semantic Differential (SD) experiment [31], the test participant have to rate a set of stimuli by a previously determined set of attributes. The attributes are, so-called Antonym Pairs (AP) (e.g. "fast" vs. "slow") and are placed on the sides of bipolar discrete 7-step scales. Each pair describes a one-dimensional feature. The result of the rating is a polarity profile for each stimulus (an example of the rating scale used in this work is given in Fig. 2.3). It shows to what extent the given features are present in the judged stimuli.

Depending on the number N of antonym pairs, one receives an N-dimensional feature space. When using the Principle Component Analysis (PCA) [32] on the averaged ratings of the test participants, the N-dimensional feature space can be reduced. The calculated components have specific Eigenvalues. All components below one (<1) are rejected, and the number of Eigenvalues above one (>1) is kept. The resulting data is a matrix where the columns represent the Principle Components, and the rows represent the coordinates of the respective AP in the dimension-reduced space. All antonym pairs influence each component, which is represented by the so-called

pixely	*					uniform
daubed		*				not daubed
shredded			*			not shredded
high contrast					*	low contrast
dismembered	*					not dismembered
jerking				*		constant
overexposed			*			underexposed
blocky			*			not blocky
flickery		*				not flickery
blurred movement	*					sharp movement
overlapped					*	not overlapped
color distorted					*	color correct
stripy				*		not stripy
blurred			*			sharp
artifical		*				natural
waggly			*			stable
noisy				*		noiseless

Fig. 2.3 Example for a polarity profile from a SD, here 17 antonym pairs were used to describe a video stimuli

Factor Loadings (FL). Now, this matrix is rotated, aiming for the maximization of the variances of the squared FL (VARIMAX criterion). In other words, the original N-dimensional feature space is approximated by another matrix with a lower rank.

The PCA is an exploratory method, and the FL have to be interpreted by the experimenter itself. The interpretation of the resulting dimensions is insofar easy since one has hints on the underlying dimensions through the grouped descriptions of the antonym pairs. The disadvantage of the method is that, since one has a limited amount of AP in the experiment, it could be possible that one misses some features simply by not asking for it. The right choice of the AP beforehand is crucial for the result. It is an effort that should not be underestimated.

2.6.2 Paired Comparison and MDS

In a Paired Comparison (PC) experiment [31], the test participants have to judge a set of stimuli. Here, pairs of stimuli are rated according to their dissimilarity. The rating scale usually has labeled ends with *very similar* to *not similar at all*. The result of this experiment is a similarity matrix for each test participant. For further data processing, the similarity matrices from all test participants are averaged.

In general, the method of Multi-Dimensional Scaling (MDS) [31] yields to find the number of dimensions required to depict a high-dimensional space in a low-dimensional space. The ratings of this test are interpreted as distances between the test stimuli. It is assumed that the perceived dissimilarities correspond to a Euclidean Metric. This means, the higher the dissimilarity, the higher the distance between the two stimuli. All distances open up a high-dimensional space. The dimensionality is depending on the number of comparisons in the test. The main aim of MDS is to find a low-dimensional space, in which the configuration can be geometrically represented while preserving the given distances. To determine the resulting dimensionality, the statistical fit parameters fed to the MDS and the ability to interpret the resulting representation are crucial. A frequently used parameter to determine if a constellation has a good fit is the STRESS. The STRESS indicates how bad the resulting distances match the given dissimilarities. Increasing the dimensionality reduces the residual error, but at the risk of introducing noise in the data. In contrast, a high number of dimensions are difficult to interpret. Therefore, it is vital to keep an eye on the interpretability. A reasonable number of dimensions can be found if the STRESS value does not increase significantly with the increasing number of dimensions. The so-called scree plots are therefore used, and ideally, a sharp *kink* can be observed, which marks the adequate number of dimensions.

This test paradigm has the advantage that the rating task is fairly simple for the test participants, and a priori provision of the stimuli features is not necessary. The disadvantage is that the interpretation of the resulting dimensions becomes only possible if the nature of the differences between the stimuli is known. In fact, the experimenter may come up with a speculative solution. Because of that, it is recommended to combine the MDS with other dimensionality minimizing methods.

2.7 Instrumental Quality Estimation

An approach to estimate the perceived quality of a transmitted media system would be quality models. Often these models are also referred to as *objective models* or *instrumental estimation*. These models use signal measurement and estimation to predict the quality rated by a person. There are several different approaches to do so, depending on the targeted use case. In this section, an overview of the central concepts of instrumental quality estimation is given.

2.7.1 Video Quality Estimation

The different models can be categorized e.g. as follows ([10, 11, 33]): First, the models can be grouped depending on the amount of information they need from the source signal.

1. Full-Reference:
   In this method, the reference signal is required as well as the transmitted signal. Here the two signals are compared, and with the help of quality metrics the

differences are quantified. If the difference is big, mostly due to impairments, the
resulting estimation of the quality rating will be low.
2. Reduced-Reference:
   Here, only some low-level features of the source signal are extracted. This information is then compared with corresponding low-level signal features of the transmitted signal. Several different proposed models using a different number of features and extracting methods estimate a quality score. To conclude, a reduced
   amount of data is necessary compared to full-reference models.
3. No-Reference:
   As the name suggests, the source signal is not needed to estimate the quality. That
   said, no comparison between source and the transmitted signal is necessary. The
   quality ratings are purely done by analyzing the transmitted signal. This bears the
   danger that actual content can be regarded as an impairment and can reduce the
   quality score (e.g. a chessboard is interpreted as block artifact).

This leads to the second way to group these models. Here, the classification is
done by the type of input the models need.

1. Planning models:
   As the name suggests, these types of models are used in the planning process, with
   the consequence that only information about the used technology is available, and
   no actual measurements.
2. Packet-header-based models:
   Here, information is taken from the packet headers. These are extracted from the
   stream and often used in monitoring processes.
3. Bitstream-based models:
   These models need the bitstream either fully or only partially decoded.
4. Signal-based models:
   Here, the decoded signal is required. These models extract information from the
   signal, e.g. the pixel information of a frame or frequency range of the audio.

There are also combined hybrid models, that, e.g. combine signal-based models
and bitstream-based models. Other combinations are also conceivable and in use.
Moreover, quality models can be categorized depending on the application they
are targeted for (e.g. streaming service, service monitoring).

2.7.2 Model Development

In the development process of a quality model, the model type is selected depending
on the target. The creation can be roughly divided into three phases: first, the selection
of the variables and parameters is taken into account; second, the building of the
model itself; third, the evaluation of the model. There should be no arguing that the
test set for the evaluation should reflect the scope of the model. Each of the mentioned
phases can then further split into details, like the choice of the modeling method, and
it depends on the needs in each case.

2.7.2.1   Feature-Based Approach to Modeling

This approach is often used by full-reference models and uses features from the source signal and the transmitted signal. Typically the two signals are compared, e.g. frame by frame or pixel by pixel. The video features could be the luminance, color value etc. For the set of features that are extracted, combined, or grouped, metrics are calculated to interpret the sequence under test.

2.7.2.2   Degradation-Based Approach to Modeling

The idea behind that approach lies in finding a metric or a set of metrics for each perceptual quality dimension, as explained in Sect. 2.5. Afterward, combining all perceptual quality dimensions estimates the overall quality. Degradations are regarded in most cases as components reducing the quality. In this strategy, variables (e.g. test conditions) will be used as input to impair the quality. The variables selected are dependent on the target application and the relevance and contribution of affecting the perceptual dimension. From this, the impact on quality is generalized for the type of input or similar input.

2.7.2.3   Subjective-Results-Based Modeling

Mapping selected variables to subjective quality ratings is here the modeling approach. This strategy is used in parametric-based quality models. The mapping is typically done by using a regression analysis, where the calculated values are aimed to match the subjective ratings. The resulting mapping function could be linear or polynomial. The goodness of fit depends on the chosen variables. The available number variables are limited to the number of parameters. The regression is often done in terms of "degradation-based modeling strategy" using the quality contribution terms related to different types of variables [10].

Model development impacting factors have been presented. The choice of the "right" approach and the "right" variables is often an iterative process. The main goal is always to predict the quality as close as possible, matching the perceived quality gathered from the test participants in the underlying experiments.

2.7.2.4   Variance Between Modeling and Subjective Ratings

All ratings obtained in subjective testing are biased. As a result, the modeled rating scores may spread widely or are distorted from one test to another. This effect is even more prominent when the test plan includes no anchor in the study. A model should not try to mimic the inter-test variation because, in the next study, the variation could be different, and a distortion of the ratings could still occur. If this shift, or distortion, between the scores is an issue, there are several methods of polynomial mapping [10] to cope with it. Here, the predicted scores are mapped to subjective scores.

2.7.3 Video Quality Metrics

Some of the most often used metrics in image and video processing are the Mean Squared Error (MSE) and the Peak Signal-to-Noise Ratio (PSNR) [11]. The MSE is the mean of the squared differences between the gray-level values of pixels in, e.g. two frames. The Root Mean Square Error (RMSE) is simply the square root of the MSE. The PSNR is a measure given in decibels and is a relative value in comparison to the maximum value a pixel can take. The main advantage of these metrics is that they are easy to compute. Nevertheless, the main drawback is that they have only a little and only approximate relationship to different kinds of degradations and quality ratings in general.

Another metric is the Structural Similarity (SSIM) index [34, 35]. It was designed as a full-reference metric to improve the abovementioned methods (MSE, PSNR). SSIM is a perception-based model that considers image degradation as a perceived change in structural information. Further, it includes information about the luminance and the contrast and their masking effects.

Even if there are further developed versions of the SSIM, the metric is related to the overall quality. Therefore, it cannot be used for the estimation of the underlying dimensions. Furthermore, in a paper, the authors state that besides its popularity, the SSIM performs "... *much closer to that of the mean squared error (MSE), than some might claim. Consequently, one is left to question the legitimacy of many of the applications of the SSIM...*" [36]. In a second paper it is stated that the SSIM is outperformed by the MSE [37] on the database containing packet-loss-impaired material.

2.7.4 Video Quality Indicators

The video quality group of the AGH University of Science and Technology[3] in Krakow developed a set of Video Quality Indicators (VQI). The general goal was to provide a set of key indicators, which describes the quality in a broad sense [38]. The selection of a subset and the usage of the indicators is dependent upon the use case they are deployed for. The latest set of indicators consists of 15 video quality indicators and 2 audio quality indicators and are publicly available [39]. The Video Quality Indicators are developed as no-reference metrics and are calculated frame

---

[3]Video quality group at AGH University of Science and Technology http://vq.kt.agh.edu.pl (last view 12/2019).

by frame. In the following list, all indicators are given. The red marked subset of the indicators is used in the later estimation of the video quality dimensions.

- Video indicators

  - Commercial Black
  - *Blockiness*
  - Block Loss
  - *Blurring*
  - *Contrast*
  - *Exposure*
  - *Flickering*
  - *Freezing*
  - Interlacing
  - Letter-boxing
  - Noise
  - Pillar-boxing
  - Slicing
  - *Spatial Activity*
  - *Temporal Activity*

- Audio indicators

  - Mute
  - Clipping

In the following a brief description of the indicators, utilized in this work, is given [38, 39].

2.7.4.1 Blockiness

This type of degradation is caused by all block-based coding techniques and results in visible blocks. The metric is calculated for pixels at boundaries of 8 × 8 blocks. Its value depends on the magnitude of color difference at the block boundary, and the contrast near boundaries. The calculated values could go to infinity, and therefore the resulting value range is limited from 0 to 3570, with higher values meaning more visible artifacts. The value range for no impairment is from 0.9 to 1.01.

2.7.4.2 Blurring

The reduction in sharpness of spatial detail and edges is called blurring—this is a result of the deleted high-frequency components during the compression process. This indicator is calculated based on the cosine of the angle between perpendicular planes in neighboring pixels. The resulting value ranges from 0 to roughly 70 with higher values meaning more visible artifacts. The value range for no impairment is from 0 to 5.

2.7.4.3 Contrast

This is a measure of the relative variation of luminance [11]. With its help, it is possible to distinguish between objects. The resulting value ranges from 0 to roundabout 120 with higher values meaning more contrast. The value range for no impairment is from 45 to 55.

2.7.4.4 Exposure

This indicator relates to the visible imbalance between the brightness of a frame. Both, a too bright and a too dark frame, are taken into account. The mean brightness of the dark and bright parts of a frame is calculated in order to detect this impairment. The resulting value ranges from 0 to 255. The value range for optimal exposure is from 115 to 125.

2.7.4.5 Flickering

The resulting value ranges from 0 to 8 with higher values meaning more visible artifacts. The indicator operates on an observation window (x consecutive frames), and the result is only given for the window, not each frame. For an 8 frame window, the typical value for no impairment is around 0.125.

2.7.4.6 Freezing

The differences in subsequent frames are investigated. In order to detect this impairment, a threshold of 80 ms is set; this is the shortest time in which a viewer can perceive this type of artifact. The resulting value ranges from 0 to 1, and the higher the value, the stronger the impairment. The value for no freezing is 0. The freezing indicator is coupled with the temporal activity indicator.

2.7.4.7 Spatial and Temporal Activity

To distinguish test- or monitored video sequences, it can be useful to compare the relative spatial information and temporal information [26]. The compression difficulty is generally related to the spatial and temporal information of a frame or Group of Pictures (GoP). Further, a deviation from the "normal" values can be an indicator of specific impairments.

The resulting value for spatial activity ranges from 0 to roughly 270, and a greater value means more spatial activity. The resulting value for temporal activity ranges from 0 to 255 and depends on the frame size. A greater value means more temporal activity.

2.7.5 Acoustic Signals: Audio Versus Speech

This section briefly discusses the terms of audio and speech.

The term audio is mainly used when it comes to the passive recording of something "audible." These are meant to be high-quality music or sound recordings. Speech is also something "audible," but here, the transmission of the spoken word or a conversation is understood. The distinction comes from the history of the development of the corresponding technology. The work of different working groups (e.g. ITU-T and ITU-R) reflects this. Audio and speech expert groups have designed different quality models for their "domain" in the past decades. However, nowadays, speech is transmitted with much better sound quality. The frequency bandwidth has been extended, for example, from narrow band to super-wide band and even full band. The applications of classical audio and speech merge more and more with the help of modern technology (e.g. videotelephony for remote participation on parties or concerts). Therefore, a discussion should be encouraged as to whether audio and speech are still separate areas. Since this work deals with videotelephony and there the voice channel is usually named audio channel, the terms audio and speech are used synonymously.

2.7.6 Audio Quality Estimation

For the instrumental audio quality estimation several quality model were developed like the ITU-T G.1070 [40], P.1201 [41], P.862 (PESQ) [42], and P.862.2 (WB-PESQ) [43]; to name a few.

The PESQ was designed to model the speech quality for narrow-band telephone networks. The later developed WB-PESQ was an extension for the usage for wideband telephone transmission. Both models are full-reference models. During the preprocessing and following some beforehand steps, time alignment between the reference or original signal and the degraded or transmitted signal is done. The central processing is a transformation of both signals in the emulation of the human auditory system. Various calculations are made (e.g. band integration, frequency, gain compensation, etc.). Finally, with the help of the calculations, a distance between the two is obtained, and subsequently, a mapping function transfers the results to an MOS [44].

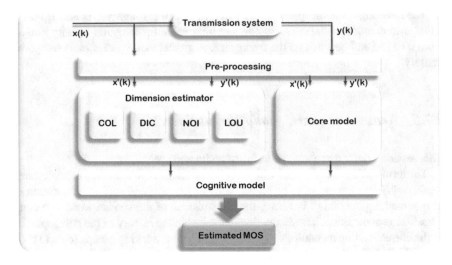

Fig. 2.4 Overview of the DIAL model framework [27]

Several other models were developed, dependent on their usage. Most of these models only give results in terms of MOS and no information about the underlying perceptual quality dimensions is obtained.

The Diagnostic Instrumental Assessment of Listening quality (DIAL) model [27] was an attempt to tackle this problem. Its framework is shown in Fig. 2.4. This full-reference model can operate with NB, WB, and SWB speech signals. It is aimed to represent the whole quality judgment process for speech signals. Its preprocessing consists first of an active speech level normalization and voice activity detection. The time alignment was taken from the PESQ model. It estimates the overall speech transmission quality and provides an MOS as a result. The necessary calculation needs 3 main steps: the *core-model* estimates nonlinear degradations from the speech processing system [44]. Further, a *dimension estimator* provides results for the four quality dimensions found in speech telephony, as described in Sect. 2.5. The last step represents a *cognitive model* that calculates scores for each perceptual dimension, which should represent the cognitive process of the human listener. The prediction ability is very high for different speech processing and transmissions systems, and it outperforms the WB-PESQ [27].

References

1. Mrak, M., Grgic, M., Kunt, M.: High-Quality Visual Experience—Creation. Processing and Interactivity of High-Resolution and High-Dimensional Video Signal. Springer, GER-Berlin (2010)
2. Sekuler, R., Blake, R.: Perception. McGraw Hill, USA, New York (2006)

3. Goldstein, E.B.: Wahrnehmungspsychologie. Spektrum Akademischer Verlag, GERHeidelberg (2002)
4. Moore, B.: An Introduction to the Psychology of Hearing, 5th edn. Elsevier Academic Press, UK, London (2007)
5. Zwicker, E., Fastl, H.: Psychoacoustics: Facts and Models. Springer, GERBerlin, Heidelberg (2007)
6. Cisco: Cisco Visual Networking Index: Forecast and Trends, 2017–2022. https://www.cisco.com/c/en/us/solutions/collateral/service-provider/visualnetworking-index-vni/white-paper-c11-741490.html. Last visit 09/2019 (2019)
7. Malaka, R., Butz, A., Hußmann, H.: Medieninformatik - Eine Einführung. Person Studium, GER-Munich (2009)
8. Belmudez, B.: Audiovisual Quality Assessment and Prediction for Videotelephony. Springer, GER-Heidelberg (2015)
9. Möller, S., Raake, A. (eds.): Quality of Experience: Advanced Concepts. Applications and Methods. Springer, GER-Heidelberg (2014)
10. Garcia, M.-N.: Parametric Packet-Based Audiovisual Quality Model for IPTV Services. Springer, GER-Berlin (2014)
11. Winkler, S.: Digital Video Quality—Vision. Models and Metrics. Wiley, UK, Chichester (2005)
12. Ghanbari, M.: Standard Codecs—Image Compression to Advanced Video Coding, 3rd edn. The Institution of Engineering and Technology, UK, London (2011)
13. Shelotkar, A.D., Barbudhe, V.K.: Video error concealment using H.264/AVC. Int. J. Comput. Sci. Inf. Technol. 3 (2012)
14. Valente, S., Dufour, C., Groliere, F., Snook, D.: An efficient error concealment implementation for MPEG-4 video streams. IEEE Trans. Consum. Electron. 47, 568–578 (2001)
15. Möller, S.: Quality Engineering - Qualität kommunikationstechnischer Systeme. Springer, GER-Heidelberg (2010)
16. Jekosch, U.: Voice and Speech Quality Perception: Assessment and Evaluation. Springer, GER-Berlin (2005)
17. Raake, A.: Speech Quality of VoIP: Assessment and Prediction. Wiley, UK, Chichester (2006)
18. von Thun, S.: Das Kommunikationsquadrat. https://www.schulz-vonthun.de/die-modelle/das-kommunikationsquadrat. Last visit 07/2019 (2019)
19. Wältermann, M.: Dimension-Based Quality Modeling of Transmitted Speech. Springer, GER-Berlin (2013)
20. Winkler, S.: Video quality and beyond. In: 15th European Signal Processing Conference (EUSIPCO). EURASIP, PL-Pozan (2007)
21. Beerends, J.G., De Caluwe, F.E.: The influence of video quality on perceived audio quality and vice versa. J. Audio Eng. Soc. 47(5), 355–362. http://www.aes.org/e-lib/browse.cfm?elib=12105 (1999)
22. ITU-T Rec. P.911: Subjective Audiovisual Quality Assessment Methods for Multimedia Applications. International Telecommunication Union, CH-Geneva. 12/1998 (1998)
23. Pinson, M., Ingram, W., Webster, A.: Audiovisual quality components. IEEE Signal Process. Mag. 28(6). IEEE—Signal Processing Society (2011)
24. Arndt, S.: Neural Correlates of Quality During Perception of Audiovisual Stimuli. Springer, GER-Berlin (2015)
25. ITU-T Rec. P.800: Methods for Subjective Determination of Transmission Quality. International Telecommunication Union, CH-Geneva (1996)
26. ITU-T Rec. P.910: Subjective Video Quality Assessment Methods for Multimedia Applications. International Telecommunication Union, CH-Geneva. 4/2008 (2008)
27. Côté, N.: Integral and Diagnostic Intrusive Prediction of Speech Quality. T-Labs Series in Telecommunication Services. Springer, GER-Berlin, Heidelberg (2011)
28. Köster, F.: Multidimensional Analysis of Conversational Telephone Speech. Springer, GER-Berlin (2018)

29. Keller, D., Seybold, T., Skowronek, J., Raake, A.: Assessing texture dimensions and video qual-
    ity in motion pictures using sensory evaluation techniques. In: 11th International Conference
    on Quality of Multimedia Experience (QoMEX). Another Medium. IEEE—Signal Processing
    Society, GER-Berlin (2019)
30. Tucker, I.: Perceptual Video Quality Dimensions. Master Thesis. Technische Universität Berlin,
    GER-Berlin (2011)
31. Bortz, J., Döringl, N.: Forschungsmethoden und Evaluation - für Human- und Sozialwis-
    senschaftler, 4th edn. Springer Medizin Verlag, GER-Heidelberg (2006)
32. Bortz, J.: Statistik - für Human- und Sozialwissenschaftler, 6th edn. Springer Medizin Verlag,
    GER-Heidelberg (2005)
33. Wu, H.R., Rao K.R. (eds.): Digital Video Image Quality and Perceptual Coding. Taylor &
    Francis Group, USA, Boca Raton (2006)
34. Wang, Z., Bovik, A.: A Universal Image Quality Index. IEEE (2002)
35. Brunet, D., Vrscay, E., Wang, Z.: On the mathematical properties of the structural similarity
    index. IEEE Trans. Image Process. **21**(2) (2012)
36. Netflix. VMAF-Video Multi-Method Assessment Fusion. https://github.com/Netflix/vmaf.
    Internet: Last view 10/2019
37. Reibman, A.R., Poole, D.: Characterizing packet-loss impairments in compressed video. In:
    2007 IEEE International Conference on Image Processing (2007)
38. Leszczuk, M., et al.: Key Indicators for Monitoring of Audiovisual Quality. IEEE, TURTrabzon
    (2014)
39. AGH University of Science and Technology, Krakow, Poland: Video Quality Indicators. http://
    vq.kt.agh.edu.pl. Internet: last view 09/2019 (2019)
40. ITU-T Rec. G.1070: Opinion Model for Video-Telephony Applications. International Telecom-
    munication Union, CH-Geneva (2012)
41. ITU-T Rec. P.1201: Parametric Non-intrusive Assessment of Audiovisual Media Streaming
    Quality. International Telecommunication Union, CH-Geneva (2012)
42. ITU-T Rec. P.862.2: Perceptual Evaluation of Speech Quality (PESQ), an Objective Method
    for End-to-End Speech Quality Assessment of Narrow-Band Telephone Networks and Speech
    Codecs. International Telecommunication Union, CH-Geneva (2001)
43. ITU-T Rec. P.862.2: Wideband Extension to Recommendation P.862 for the Assessment of
    Wideband Telephone Networks and Speech Codecs. International Telecommunication Union,
    CH-Geneva (2007)
44. Hinterleitner, F.: Quality of Synthetic Speech—Perceptual Dimensions, Influencing Factors,
    and Instrumental Assessment. T-Labs Series in Telecommunication Services. Springer, GER-
    Berlin, Heidelberg (2017)

Chapter 3
Quality Feature Space of Transmitted Video

In this chapter, the quality-relevant dimensions of transmitted video are identified. The results are obtained from exploratory experiments. This chapter gives a brief explanation of the used methods and a detailed description of the conducted experiments.

3.1   Experimental Setup

In this section, the experimental setup is described. First, details on test conditions are given. The different steps of the processing chain are explained, and all test conditions and degradations are briefly characterized. Second, an insight into the test rooms and the equipment in use is provided. Furthermore, an overview of all test participants in this work is presented.

In the following list, the acronyms for the subjective tests and brief descriptions are given. These acronyms are used from here on.

- **SD-PCA—Test**: Semantic Differential (SD) test with subsequent Principle Component Analysis (PCA).
- **PC-MDS—Test**: Paired Comparison (PC) test with subsequent Multi-Dimensional Scaling (MDS).
- **VQDIM I—Test**: Direct scaling of the video quality dimensions with head-and-shoulder video material.
- **VQDIM II—Test**: Dimension measuring—direct scaling of the video quality dimensions, to obtain an insight into the scale range for each dimension.

© The Author(s), under exclusive license to Springer Nature Switzerland AG 2021
F. Schiffner, *Dimension-Based Quality Analysis and Prediction
for Videotelephony*, T-Labs Series in Telecommunication Services,
https://doi.org/10.1007/978-3-030-56570-1_3

Table 3.1 Overview of the source material

Material	Videotelephony/head-and-shoulder scene
File duration	Approx. 10 s
Resolution	640×480 pixel
Frame rate	25 fps
Audio sampling rate	$Fs = 16$ kHz
Quantization	Bit depth = 8 bit
Audio level	-26 dBov
Language	Native German speech
Scenario I	Kitchen purchase (female)—living room
Scenario II	Doctors appointment (female)—office
Scenario III	Car reservation (male)—office
Scenario IV	Birthday party (male)—living room

- **VQDIM III—Test**: Direct scaling of the video quality dimensions with broad video content.
- **AV-DIM—Test**: Direct scaling of the audio and video quality dimensions with head-and-shoulder video material.

3.1.1 Processing of the Test Material

3.1.1.1 Source Material

The source material was taken from a video database produced by Belmudez [1]. These video sequences are short scenarios of simulated videotelephony conversations and aimed to reflect everyday communication. Therefore, content like a "doctors call," or a "car reservation" was chosen. In Fig. 3.1, examples for the used head-and-shoulder sequences are shown. In order to make the video more "intention attracting," visual cues like body gestures or pointing on a calendar were added by the creator. In Table 3.1, an overview of the source material is given. For each scenario, 10 short clips were recorded with a duration of approximately 10 s each. Two female and two male speakers with two different background settings were used. One setting is office like, and the other one is meant to simulate a living room. The recordings were in raw format with VGA resolution and 25 frames/s frame rate. The audio channel was recorded with a sampling frequency of 16 kHz and 8-bit quantization.

3.1.1.2 Video Degradations and Processing Chains

The test material in this work was prepared to cover a broad range of potential video degradations in videotelephony. The degradations were selected on the basis of an expert survey[1] and mainly focused on potential transmission degradations. In the following, the creation of the test material is described in detail. Additional information on the degradations can be found in Table 3.2. Examples of the test material are shown in Fig. 3.1.

To introduce degradations into the video, the *Reference Impairment System for Video (RISV)* [2] was employed. The RISV is an adjustable system that can be used to create reference conditions and can produce video degradations that can occur in digital video systems. To generate most types of degradations, a *MATLAB* [4] processing chain was built. The host script and examples of the impairment processing are shown in Listing C.1 and C.2.

Blurring. This impairment is defined as a reduction in the sharpness of the edges and results in lesser spatial detail. This degradation is often caused by compression algorithms, as a result of a trade-off between bits to code the video resolution and the motion. For including blurring, the ITU-T Rec. P.930 [2] recommend a symmetric, one-dimensional, 15-tap Finite Duration Impulse Response (FIR) low-pass filter. A processing chain was built in *MATLAB*. Exemplary for the video file processing, the code for that degradation can be seen in Listing C.2. For a moderate impairment filter setup 1 was chosen. To include a strong impairment, an additional filter ("Filter7") was created. The filter coefficients can be seen in the code (comp. Listing C.2). In both cases, the filter was applied line-wise.

Blockiness. This impairment also known as block distortion refers to a block-like pattern in a picture. It is caused often by too big quantization steps during the compression of individual blocks (typically 8×8 pixel in size). This leads to discontinuities at transitions between two blocks and to a loss of high-frequency components, which results in visible square blocks.

Based on the description in ITU-T Rec. P.930 [2], the video was read frame by frame. An error pattern was created dependent on block size and the number of blocks that should be impaired for each frame. Following, the values for each color layer were averaged, and all blocks and frames were added to the final video.

Jerkiness. It is often caused due to waiting too long at a transmission knot. The receiver is waiting for the next frame and holding the before received. Therefore, it can be described as a disintegration of a continuous flow into a series of distinct single frames. The effect was produced by holding a single frame over a certain number of frames (e.g., 3 or 6 frames).

Noise Insertion. To introduce noise in the video, the material was transformed from the RGB colorspace into YCbCr colorspace. Following this, only the luminance values were impaired. For every frame, a new random error pattern was created. The

---

[1]More details on the survey in Sect. 3.2.1.

error pattern indicates which pixel should be affected. The number of impaired pixels is set before in % dependent on the number of pixels in one frame. The luminance values of the affected pixels were set to random values.

Luminosity Impairment. If the luminosity of a video is impaired, it expresses itself as too bright or too dark. Overexposure is also referred to as "washed out" since the bright parts of the image lost their details. In contrast, underexposure lost details in the shadow parts. These parts are difficult or not distinguishable from black. An impaired luminosity is not a typical transmission degradation, but a degradation, resulting from a wrong setup in video recording or a malfunctioning device. Since it is regarded as vital by the expert survey when regarding video quality, this impairment type is included in this work.

To produce over- and underexposure, the luminance value of each pixel in each frame was raised or reduced by a fixed value. In this manner, an equal shift of the luminosity to a darker or lighter appearance was achieved.

Coding Artifacts. Artifacts of coding can be introduced in the video during the transcoding process from one codec to another. This is done to enable interoperability between different networks and systems. Also, in certain circumstances, the network may decide to send a series of packets, which are encoded with a lower bit rate when there are transmission limitations. This results in a noticeable drop in quality for a period of time until enough transmission bandwidth is available.

To introduce effects of coding in the data set, the material was encoded with different bit rates using the H.264 video codec. To ensure the same level of degradation over the whole sample, 2-pass encoding was realized using *ffmpeg* [3] called from *MATLAB* [4].

Packet Loss Artifacts. This impairment occurs when one or more packetized data units (packets) fail to arrive at their destination when traveling across a network. This can be due to discarded packets in the router when the packets are arriving too late, or it is too busy to deliver the packets in an appropriate time.

Packet loss can express itself in different ways, depending on the number of packets lost and the algorithms in the encoder dealing with packet loss (Packet Loss Concealment). Packet loss is often visible as a partial or complete "green out" of the image, as well as missing slices or grouped block artifacts. Often only some parts of an image are impaired, while the rest remains intact (comp. Fig. 3.1, No. 8). The degradations were processed via *FFmpeg*, *Netem* [5], and *Traffic Control* and a Linux OS was, therefore, used. The video files were sent through a network simulator (*Netem*) via *FFmpeg* and recorded with *FFmpeg* afterward. In the network simulator, packet loss was forced by setting a loss rate via *Traffic Control*. The packet loss could occur randomly over the whole video sample and no bursty packet loss was included. The *FFmpeg* commands can be seen in Listing C.4.

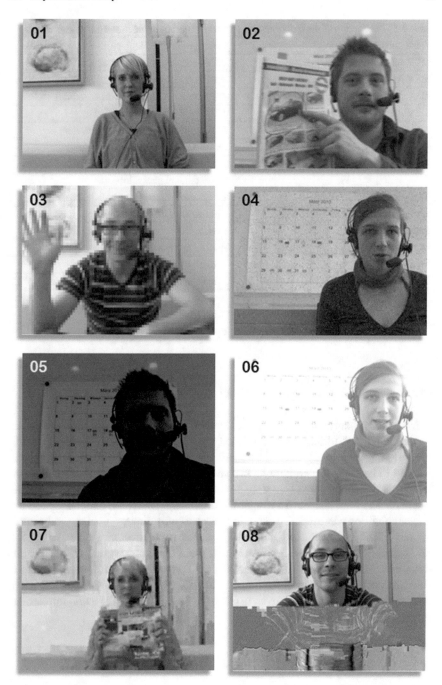

Fig. 3.1 Examples of the video material—01 Reference; 02 Artificial Blurring; 03 Artificial Block-iness; 04 Artificial NoiseQ; 05 Luminance Impairment I (darker); 06 Luminance Impairment II (lighter); 07 Coding Artifacts (H.264); 08 Packet Loss

Table 3.2 Description of the impairments in the test material for video and audio impairments

Video impairment	Description
RISV Artificial Blockiness $A \times A$	All frames impaired ($A \times A$ pixel block size)
RISV Artificial Blurring ITU(F1–F6)	All frames impaired (filter setting 1–6 from Rec.)
RISV Artificial Blurring Filter7	All frames impaired (own filter setting)
RISV Artificial Jerkiness X Frames	Jerkiness (X frames held)
RISV Artificial NoiseQ $X\%$	Salt & Pepper noise ($X\%$ pixel/frame)
H.264 Bitrate xx kbps	H.264-Codec bitrate (2-pass coding)
Packet Loss $x.x\%$	H.264-Codec, Traffic Control, NetEm
	$x.x\%$ random packet loss rate
Luminance Impairment I (darker)	Luminance value reduced (underexposure)
Luminance Impairment II (lighter)	Luminance value raised (overexposure)
Audio impairment	Description
Level -16dBov	$+10$ dB speech level raise
Level -36dBov	-10 dB speech level reduction
G.722.2 (PL0.2% & PL0.8%)	Wide band codec (random packet loss rate)
G.711	Narrow band codec
G.722.2 (6.6 kbits)	Wide band codec (low bitrate)
MNRU-45 & 55	Speech modulated noise (2 settings)

3.1.1.3 Audio Degradations and Processing Chain

To introduce impairments in the audio channel, the *Software Tool Library* provided by the ITU-T was utilized [6], and called via *MATLAB* scripts. The choice for the different test conditions for the audio channel is inspired by the work from [7, 8]. The aim was to have the test conditions that should be able to trigger the corresponding speech quality dimensions. Therefore, two test conditions per perceptual speech quality dimension were chosen. To trigger the dimension *Loudness*, one louder and one quieter condition was used. To impair the *Discontinuity* dimension of the audio channel, the *G.722.2* [9] was used and random packet loss introduced. The two codecs, *G.711* [10] and *G.722.2* with a low bitrate were chosen, to affect the *Coloration* dimension. For introducing noise into the audio, the *MNRU* is used as described in [6]. More details on the test conditions and the perceptual speech quality dimensions in an audio-visual context are given in Chap. 7. An example of the audio impairment processing is given in Listing C.3.

3.1.2 Test Rooms

In this section, a description of the test rooms is given. All subjective experiments were conducted in rooms, which qualified in their properties as a test location, according to ITU-T Rec. P.910 [11]. These rooms were equipped with an office chair and a table. There the monitor and the test laptop were set up. In the experiments with an active audio channel, an audio interface and headphones were also set up. All equipment used in the studies is listed in Table 3.3. An example of the test room setup is shown in Fig. 3.2. The artificial light was set in a manner that provided sufficient ambient light and made sure not to be overly bright to constitute a distraction. Natural light from the outside was blocked to ensure that the light condition would not change throughout the experiment duration.

3.1.3 Test Participants

In this section, an overview of all test participants that took part in the subjective video quality testings is given in Table 3.4. The participants were recruited mainly with the help of an online test participants database hosted by the Technische Universität

Fig. 3.2 Example of the test setup—here VQDIM II test setup

Table 3.3 Overview of the test equipment used for the subjective testing

Test laptop I	Sony Vaio PCG-81312M (SD-PCA, PC-MDS-tests)
Test laptop II	Fujitsu Lifebook S761 (VQDIM I-III, AV-DIM-tests)
Operating system	Window 7 Professional 64bit
Monitor	Viewsonic VP2650wb (Res 1920 × 1200)
Audio interface	Roland Edirol UA25EX
Head phones	AKG K601
Controls	Logitech M100 optical mouse

Table 3.4 Overview of all test participants in this work. Number of participants, gender, age, and standard deviation

Gender	SD-PCA			PC-MDS			VQDIM I		
	No.	Age	σ	No.	Age	σ	No.	Age	σ
Female	11	35.2	7.3	11	29.6	6.1	15	29.5	8.2
Male	12	32.4	6.6	15	30.5	6.3	13	28.2	6.1
Total	23	33.9	7.1	26	29.9	6.4	28	28.9	7.3
	VQDIM II			VQDIM III			AV-DIM		
Female	19	29.7	3.9	13	30.9	6.5	17	31.2	7.7
Male	28	28.7	5.1	12	31.5	6.6	15	33.9	10.7
Total	47	29.1	4.7	25	30.8	6.5	32	32.5	9.3

Berlin[2] (Berlin Institute of Technology). The participants in this database reflect a broad range of the German population—there are no restrictions to subscribe to the database. The database members voluntarily choose every test participation.

All test participants in this work received compensation for there participation. Every participant was subjected to a vision test (*Ishihara Test* [12], *Snellen Table* [13]) before the experiment to check for normal eyesight. An example of the vision test can be seen in Fig. 3.3. The minimum requirement was to recognize five different probes from the Ishihara Test and to reach an acuity in the Snellen Chart of a minimum of 20/25 (line No. 7). The participants were not audio-metrically screened. All participants stated that no hearing impairment was known to them. Furthermore, in the AV-DIM experiment, the test participants were allowed to adjust in the beginning the level accordingly to the individual MCL (*Most Comfort Level*).

---

[2]"Probandenportal der Technischen Universität Berlin"—https://proband.ipa.tu-berlin.de/.

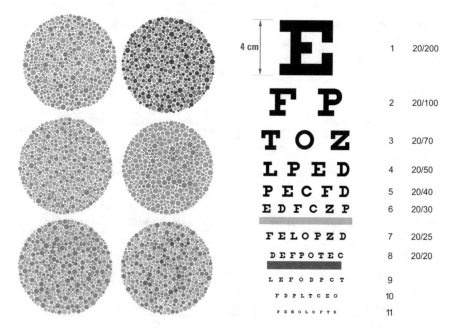

Fig. 3.3 (left) Examples form the Ishihara Test [12]; (right) Snellen Table [13]

3.2 Semantic Differential—Test

3.2.1 Determination of the SD Attributes

For the SD test paradigm, a given predefined set of attributes is needed. These attributes are presented in pairs on scales with the opposing attributes on each side, as described in Sect. 2.6.1. These word pairs are called antonym pairs, and each represents a one-dimensional feature.

In order to find the meaningful attributes in terms of video quality, three pretests were conducted. To start with, an expert survey was conducted. The survey was split into two tasks. First, 11 experts in the field of video coding, video quality, and quality assessment were asked to name attributes that describe typical video degradations. The participants use German or English. The terms described were adjectives (e.g."natürlich—naturally"), nouns (e.g."Natürlichkeit—Naturalness"), or antonym pairs (e.g."natürlich–unnatürlich vs. naturally–unnaturally"). If needed, the participants could write a self-created description. Second, a list of antonym pairs was prepared based on the work from [14]. The experts were asked to chose 10 antonym pairs from a list of 26, to see which suited best as a description for video impairments. The questionnaire for the expert survey can be viewed in Appendix A. This was done without watching any video to avoid biased wording from the appearance of showed video material. Afterward, all given attributes were collected, grouped,

Table 3.5 List of the antonym pairs—as finally resulted from the pretests (in brackets the original German attributes)

Adjective	$\Longleftrightarrow$	Antonym
Pixely (verpixelt)	$\Longleftrightarrow$	Uniform (einheitlich)
Daubed (matschig)	$\Longleftrightarrow$	Not daubed (nicht matschig)
Shredded (zerfasert)	$\Longleftrightarrow$	Not shredded (nicht zerfasert)
High contrast (kontraststark)	$\Longleftrightarrow$	Low contrast (kontrastschwach)
Dismembered (zerstückelt)	$\Longleftrightarrow$	Not dismembered (zusammenhängend)
Jerking (ruckartig)	$\Longleftrightarrow$	Constant (flüssig)
Overexposed (überbelichtet)	$\Longleftrightarrow$	Underexposed (unterbelichted)
Blocky (blockig)	$\Longleftrightarrow$	Not blocky (nicht blockig)
Flickery (flimmernd)	$\Longleftrightarrow$	Not flickery (nicht flimmernd)
Blurred movement (bewegungsunscharf)	$\Longleftrightarrow$	Sharp movement (bewegungsscharf)
Overlapped (überlagert)	$\Longleftrightarrow$	Not overlapped (nicht überlagert)
Color distorted (farbverzerrt)	$\Longleftrightarrow$	Color correct (farbrichtig)
Stripy (streifig)	$\Longleftrightarrow$	Not stripy (ungestreift)
Blurred (unscharf)	$\Longleftrightarrow$	Sharp (scharf)
Artificial (künstlich)	$\Longleftrightarrow$	Natural (natrülich)
Waggly (wackelig)	$\Longleftrightarrow$	Stable (fest)
Noisy (verrauscht)	$\Longleftrightarrow$	Noiseless (rauschfrei)

and translated to German when necessary. The most suitable terms were chosen for the second pretest.

In the second pretest, a prepared set of impaired stimuli was shown to a group of 8 experts. The task was to choose all attribute pairs that could describe the stimuli. Additionally, the experts had the possibility to describe the stimuli if there are more or missing attributes to describe the video (comp. Appendix A). As a result, the most frequently selected pairs were chosen, and the final set of 17 antonym pairs (see Table 3.5) was used in the main SD test.

All pairs, besides the two "high contrast versus low contrast" and "overexposed versus underexposed," have the attribute on the right side, referring to "no degradation." These two pairs have their optimum value in the middle of the scale. Otherwise, these scales would have to be split up into two separate scales each. That would mean to lengthen the SD without any additional benefit. In the third pretest, 6 experts from the field of quality assessment are invited. The task was to watch a set of degraded test files and decide which of the antonym pairs can be observed in the video samples. This was done to make sure that the set of antonym pairs actually reflects the impairments. Through this, the experts can verify the feasibility of the planned SD study. The results from the third pretest were promising, and the complete study was enrolled (see Sect. 3.2.2).

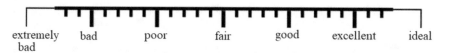

extremely bad   bad   poor   fair   good   excellent   ideal

Fig. 3.4 7-point continuous scale, labels shown here are translated from German to English (corresponding values and German scale labels in brackets): extremely bad (1/extrem schlecht), bad (2/schlecht), poor (3/dürftig), fair (4/ordentlich), good (5/gut), excellent (6/ausgezeichnet), and ideal (7/ideal)

3.2.2 SD Test—Test Procedure

The test was conducted with naïve participants (see Table 3.4). The duration of the experiment was between 40 and 60 min in total, including greeting, introduction, consent form, and vision test. The test was located in test rooms specified in Sect. 3.1.2. In the beginning, a small training was placed to allow the participants to get familiar with the rating task and the degradations. The participants were allowed to take a 5 min break if needed in the middle of the session. The degradations used in this test are listed in Table 3.6. The first task was to rate the overall quality of the video samples via a 7-point continuous scale (Fig. 3.4) [15].

The second task was to describe the video via a *SD*. The set of antonym pairs described in Sect. 3.2.1 was used to get a polarity profile for each degradation. The order of the antonym pairs was randomized for each test person. The same was done for the test file play order. Between one antonym pair, a discrete 7-step scale was placed. The participants had to weigh which one of the two words framing the scales describes the sample best. A modified version of a browser-based questionnaire (TheFragebogen [16]) was used as user interface for the *SD*. The interface is shown in Fig. 3.5.

3.2.3 Test Results and Interpretation

The results of the conducted *SD* experiment were analyzed. First, the overall quality ratings are regarded. Here it was found, as one would expect, the bigger the degradation, the smaller the perceived quality (see Table 3.6 and Fig. 3.6). The standard deviation lies within the range of standard deviations typically found in subjective quality testings. In comparison to a previous audio-visual quality study [17], the results show a very strong correlation ($r = 0.97$) and a low RMSE of 0.21. The overall video quality ratings from the two studies are compared in Fig. 3.7.

The polarity profiles obtained in the SD were analyzed, and a PCA was conducted. The rotation method was VARIMAX with Kaiser normalization. Further, KMO and Bartlett's tests are used to check whether the data is good enough for factor analysis. The resulting values are, for KMO .67, and for Bartlett test $p < 0.001$, indicating that the data is good enough. The result of the PCA reveals 4 components with

Beschreiben Sie bitte die Probe unter zur Hilfenahme der angegebenen Gegensatzpaare.
Bitte setzen Sie bei allen Paaren ein Kreuz in das jeweilige Kästchen.

verpixelt				*				einheitlich
matschig			*					nicht matschig
zerfasert				*				nicht zerfasert
kontraststark						*		kontrastschwach
zerstückelt	*							zusammenhängend
ruckartig				*				flüssig
überbelichtet			*					unterbelichtet
blockig				*				nicht blockig
flimmernd			*					nicht flimmernd
bewegungsunscharf					*			bewegungsscharf
überlagert						*		nicht überlagert
farbverzerrt							*	farbrichtig
streifig		*						ungestreift
unscharf				*				scharf
künstlich		*						natürlich
wackelig						*		fest
verrauscht			*					rauschfrei

Replay video

Next

Fig. 3.5 GUI used to gather the polarity profiles in the SD experiment. Translation of the antonym pairs can be found in Table 3.5. Translation of the instructions on top: "Please describe the sample with the help of the indicated antonym pairs. Please tick a respective box for every antonym pair"

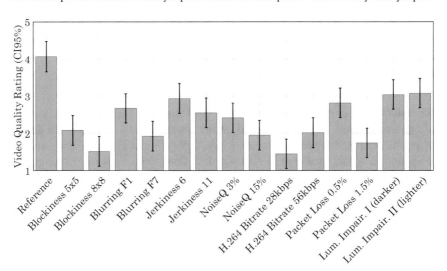

Fig. 3.6 Video quality rating with confidence interval (CI95 %)

Table 3.6 Video quality (VQ) rating from the SD experiment with 95% Confidence interval (CI95%)

Condition	VQ	CI95%
Reference	4.07	0.41
Blockiness 5 × 5	2.09	0.40
Blockiness 8 × 8	1.52	0.40
Blurring F1	2.67	0.40
Blurring F7	1.93	0.40
Jerkiness 6	2.94	0.40
Jerkiness 11	2.56	0.40
NoiseQ 3%	2.42	0.40
NoiseQ 15%	1.95	0.40
H.264 Bitrate 28 kbps	1.45	0.39
H.264 Bitrate 56 kbps	2.01	0.40
Packet Loss 0.5%	2.82	0.40
Packet Loss 1.5%	1.74	0.39
Lum. Impair. I (darker)	3.05	0.40
Lum. Impair. II (lighter)	3.08	0.40

Fig. 3.7 Comparison of the quality ratings form the SD-test and AV-test. For comparability scores are normalized (0–1)

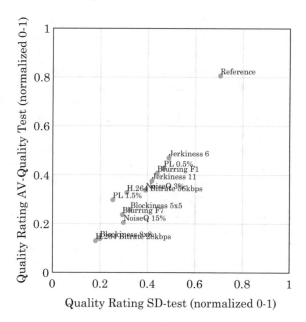

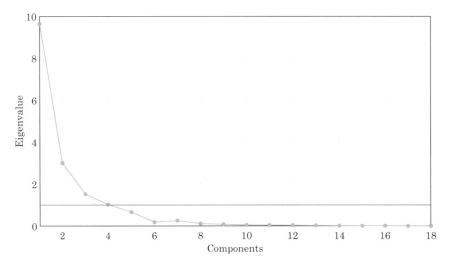

Fig. 3.8 Scree plot for the PCA on the SD experiment

Eigenvalues above 1 (see columns 4–7 in Table 3.7). To illustrate that, a scree plot is given in Fig. 3.8.

To interpret which antonym pairs load on which component, only factor loadings above 0.7 are taken into account. Component 1 is loaded by the 7 antonym pairs (1, 2, 4, 12, 13, 14, 15) and explains 56.7% of the variance. The component is labeled *Unclearness* since the describing antonym pairs are related to an unclear video image. This dimension has temporal and spatial aspects. Component 2 explains 17.7% of the variance and is loaded by pairs 5, 6, 8, 10, 16. The antonym pairs describing impairments are related to a broken and incomplete video, and therefore labeled as *Incompleteness*. This dimension seems to have more temporal aspects. The component 3 is loaded by pairs 9, 17, and explains 8.9% of the variance. The label for that is *Noisiness* and only related to the inserted noise. The component 4 is loaded only by pair 7 and is labeled *Luminosity*. It can be linked to both luminance impairments and explains 6.0% of the variance. It is not possible to clearly distinguish between *spatial* and *temporal* dimensions from the data. The two antonym pairs (3,11) could not be assigned to one of the four described dimensions since the factor loadings are always lower than 0.7. No clear distinction between the components could be furthermore observed. The resulting perceptual video quality space after the *SD* experiment is summarized in Table 3.8.

Table 3.7 Columns 2 and 3 show the antonym pairs (only English translation); column 4–7 show the factor loading (≥0.7) of the antonym pairs on the components (CP) with Eigenvalues above 1 (last row)

	Adjective	Antonym	CP 1	CP 2	CP 3	CP 4
1	Pixely	Uniform	0.70			
2	Daubed	Not daubed	0.84			
3	Shredded	Not shredded				
4	High contrast	Low contrast	−0.86			
5	Dismembered	Not dismembered		0.97		
6	Jerking	Constant		0.96		
7	Overexposed	Underexposed				0.99
8	Blocky	Not blocky		0.76		
9	Flickery	Not flickery			0.88	
10	Blurred movement	Sharp movement		0.70		
11	Overlapped	Not overlapped				
12	Color distorted	Color correct	0.80			
13	Stripy	Not stripy	0.70			
14	Blurred	Sharp	0.87			
15	Artificial	Natural	0.72			
16	Waggly	Stable		0.90		
17	Noisy	Noiseless			0.96	
		Eigenvalues	9.64	3.01	1.52	1.02

Table 3.8 Resulting perceptual video quality dimensions derived from the SD experiment

	Name	Explained variance (%)	Explained variance (cumulative) (%)
CP 1-Dimension I	Unclearness	57.16	57.16
CP 2-Dimension II	Incompleteness	18.38	75.54
CP 3-Dimension III	Noisiness	9.21	84.76
CP 4-Dimension IV	Luminosity	6.02	90.77

3.3 Paired Comparison—Test

3.3.1 PC Test—Procedure

As the name of the method already suggests, a pair of stimuli will be compared. Having N conditions in the experiment is leading to a number of Num $= N(N − 1)$ comparisons. Taking into account that also each condition is compared with itself,

Table 3.9 PC test—test conditions

Name	Description
Reference	Unimpaired material
Bitrate 28 k	H.264 Bitrate 28 kbps
Bitrate 56 k	H.264 Bitrate 56 kbps
Blocki8 $\times$ 8	RISV Artificial Blockiness 8 $\times$ 8
Blocki5 $\times$ 5	RISV Artificial Blockiness 5 $\times$ 5
Blurr$F$7	RISV Artificial Blurring Filter7
Blurr$F$1	RISV Artificial Blurring ITU (Filter 1)
LumValdark	Luminance Impairment I
LumVallight	Luminance Impairment II
Jerki11Frame	RISV Artificial Jerkiness 11 frames
Jerki6Frame	RISV Artificial Jerkiness 6 frames
NoiseQ15	RISV Artificial NoiseQ 15%
NoiseQ3	RISV Artificial NoiseQ 3%
pl05	Packet Loss 0.5%
pl15	Packet Loss 1.5%

it is needed to add N comparisons to the test set. It is assumed that the similarity rating would be the same, whether "A" is compared with "B" or "B" is compared with "A." Therefore, the overall test set of 240 comparisons is split into half. It was chosen to have all comparisons in the final analysis to create a complete distance matrix for the following MDS. Therefore, one test person had the rate the "upper-half" of the similarity matrix and one the "lower-half," resulting in 119 comparisons for one test session. The order of the comparison was alternated between the test persons. This means, the first participant had to rate the similarity between "A" and "B" ("upper-half"), and the second participant had to rate between "B" and "A" ("lower-half"). Furthermore, the order of the video pairs was randomized for each participant. An equally distributed number of female and male speakers were chosen from the source material and were equally distributed in the comparisons. In Table 3.9, the test conditions are given. More details on the test conditions are provided in Sect. 3.1.1.

The test was conducted with naïve participants (see Table 3.4). The duration of the experiment was approximately 50 min, including greeting, introduction, consent form, and vision test. The test was carried out in a room according to [11] as in the SD experiment before (see Sect. 3.1.2). The participants were allowed to take a 5 min break if needed in the middle of the session. The task was to rate the similarity of two subsequently presented video samples via an 11-point continuous scale (see ITU-T P.910 [11]). In between the two video samples, a gray screen was shown for 2 s. On the left side of the rating scale, the corresponding values were placed, and on the right side, the attributes ("very similar—sehr ähnlich—not similar at all—gar nicht ähnlich") to give the participants guidance when rating the similarities. The

Fig. 3.9 GUI from the test tool used in the PC experiment. Translation of the instructions on top: "Evaluation of the similarity of the samples"

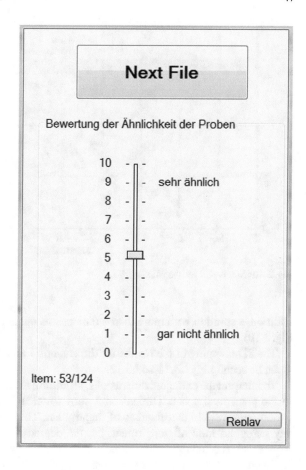

user interface[3] that was used in the experiment is shown in Fig. 3.9. A training run, consisting of five test files, was placed at the beginning to allow the participants to get familiar with the rating task and the degradations.

3.3.2 Test Results and Interpretation

The scores of the paired comparison were analyzed, and a classical MDS was calculated. A 4-dimensional constellation was chosen. The Kruskal Stress-1 was 0.057 for the 4 dimensional solution—this can be regarded as *good*. A solution with 3 dimensions was not sufficient since the Kruskal Stress-1 had a value of 0.079. In

---

[3]The used test tool named *AVrate* was developed by the TU Berlin and Telekom Innovation Laboratories 2007, last update 2018 [18].

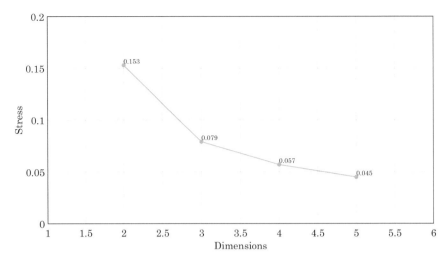

Fig. 3.10 Stress plot for the MDS

addition, a solution with more than 4 dimensions did not provide a better fit (comp. Fig. 3.10).

The 2D mapping of the results for dimensions 1 and 2 and for dimensions 3 and 4 can be seen in Figs. 3.11 and 3.12.

To interpret the extracted dimensions, all stimuli are investigated for their position in the perceptual space. Dimension 1 shows a clear tendency from the *reference* condition toward higher degrees of impairment. This means it cannot be linked to any particular kind of impairment, but the degradation in general. It can also be interpreted as a "quality dimension."

To further prove this interpretation, the positions of the test condition on this dimension were compared with the quality ratings from the SD test (comp. Sect. 3.2.3). Hence, the position values and the quality scores are normalized, and the PEARSON correlation was calculated. The result is shown in Table 3.10. The correlation coefficient is $r = -0.82$ and indicates a strong negative linear relationship. The negative value comes from the orientation of dimension 1. Because, when calculating a MDS, the algorithm does not care about the orientation, only the relationship and the distances are of importance. Resulting from that investigation dimension 1 is labeled *degree of degradation*. The dimension 2 shows a spreading from *noise to no noise* with a clear gap observed between the two noise conditions and the other test conditions. This is interpreted as a noise dimension, and therefore labeled as *Noisiness*. The label of the dimension 3 is not that easy to interpret as it seems to refer to a blurred and unclear image. Both blur-test-conditions are on the far end of the axis, with a big gap to the other test conditions. However, there are several contradictions in the positions of the different test conditions e.g. *bitrate56k* or *noiseQ15*. The dimension 3 is labeled *Unsharpness*. Similar to dimension 2 and 3, the dimension 4 shows a separation for *optimal to suboptimal luminosity*, but not so clearly. It was expected,

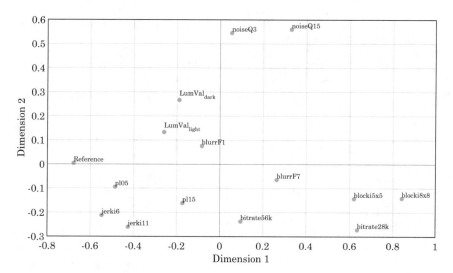

Fig. 3.11 MDS-test result—"Dimension 1 versus Dimension 2"

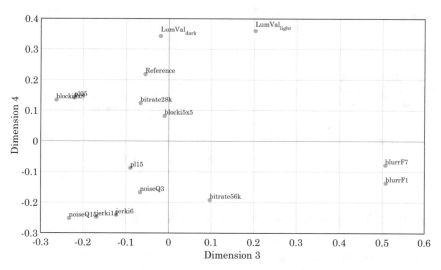

Fig. 3.12 MDS-test result—"Dimension 3 versus Dimension 4"

that the other test condition group closer together, instead they spread along the dimension axis. The dimension is labeled as *Suboptimal Luminosity*. Summarizing, the labeled perceptual dimensions are listed in Table 3.11.

Table 3.10 Comparison of the dimension 1 (DIM 1) with the quality ratings from the SD test ($SD\ Q_{\text{rate}}$), Values are normalized [0–1]

Name	DIM 1	$SD\ Q_{\text{rate}}$
Reference	0.00	1.00
Bitrate 28 k	0.86	0.00
Bitrate 56 k	0.51	0.21
Blocki8 × 8	1.00	0.03
Blocki5 × 5	0.85	0.24
Blurr$F7$	0.62	0.18
Blurr$F1$	0.39	0.47
LumValdark	0.32	0.61
LumVallight	0.27	0.62
Jerki11Frame	0.17	0.42
Jerki6Frame	0.09	0.57
NoiseQ15	0.66	0.19
NoiseQ3	0.48	0.37
$pl05$	0.13	0.52
$pl15$	0.33	0.11
Pearson R		−0.82

Table 3.11 Resulting perceptual video quality dimensions derived from the PC test

	Name
MDS-Dimension I	Degree of degradation
MDS-Dimension II	Noisiness
MDS-Dimension III	Unsharpness
MDS-Dimension IV	Suboptimal Luminosity

3.4 Synthesis—Perceptual Video Quality Space

In this section, the results from previous experiments (SD test, PC test, and the work from Tucker [14]) are compared. The analysis and the final conclusion are provided.

The interpretation of the resulting scatter plots (Figs. 3.11 and 3.12) from the PC-MDS test, for each pair of dimensions, unveils that the less weightful dimensions *Noisiness* and *Luminosity* from the SD-PCA test can also be observed. The first and third PC-MDS-dimensions cannot be clearly translated to the results of the SD-PCA-test. For the first PC-MDS-dimension, a label from the SD-PCA experiment [19] cannot be applied, because there is not enough overlap. This dimension shows an obvious connection to the intensity of the included impairments. Therefore, it is argued that this dimension appears as some overall-quality-related dimension where the reference condition is placed on one end of the scale, and the higher the impairment, the greater the distance from that reference condition. As a result, one

Table 3.12 Result of the second PCA for the SD-PCA-dimension *Unclearness*. Columns 1 and 2 show the antonym pairs; columns 3 and 4 show the factor loading of the antonym pairs on the components (CP)

Attribute	Antonym	CP 1	CP 2
Blocky	Not blocky	0.94	
Color distorted	Color correct		0.69
High contrast	Low contrast		-0.99
Artificial	Natural	0.64	0.63
Daubed	Not daubed	0.64	0.67
Stripy	Not stripy	0.88	
Blurred	Sharp		0.73
Pixely	Uniform	0.82	
Shredded	Not shredded	0.92	
Overlapped	Not overlapped	0.87	

cannot use this dimension to decompose the overall quality in its underlying dimensions. The results for the third PC-MDS-dimension *Unsharpness* show a tendency that could be mapped to the SD-PCA-dimension *Unclearness*. In the SD-PCA test, attributes like "blurred," "daubed," and "low contrast" were loaded the most on this component, which can be directly linked to the "blurred" test conditions in the PC-MDS test. The PC-MDS test results are not entirely the same as the results from the SD-PCA-test. However, this gives supporting information to set up a perceptual space for video quality. The results from the SD-PCA, PC-MDS, and Tucker were taken into account since these works support and supplement each other. In the work from Tucker [14], two spatial dimensions (*Fragmentation, Freq. Content*) and one temporal dimension (*Movement Disturbance*) are presented.

Since lots of antonym pairs were loaded on the SD-PCA-dimension *Unclearness* and show mostly spatial effects, an additional PCA was conducted (see Table 3.12). This was done to investigate if the two spatial dimensions from Tucker can be found there. In fact, the result unveiled the two dimensions relating to spatial impairments. The result can be divided into "a broken, incomplete, stripped" image on the one hand (see marked blue results in Table 3.12) and in "smeared, blurred, unclear, color distorted" on the other hand (see magenta marked results in Table 3.12). This matches entirely with the two spatial dimensions from Tucker. Also, the antonym pairs "artificial versus natural" and "daubed versus not daubed" (marked in green in Table 3.12) are present in both components to the same degree. Therefore, they cannot be used to distinguish between the two spatial dimensions. This is completely understandable since, e.g."artificial" could describe both, a "stripy" video or a "color distorted" video.

In view of the test results, the following interpretations and considerations, the final quality-relevant perceptual space for video in the domain of videotelephony consists of five dimensions. The first dimension, namely *Fragmentation*, refers to a "fallen

Table 3.13 Summary of all test results (columns 1–3); proposed perceptual quality dimensions for video (VQD) (column 4)

SD-PCA	PC-MDS	Tucker	VQD
Unclearness	—	Fragmentation	I Fragmentation
—	Unsharpness	Freq. Content	II Unclearness
Incompleteness	—	Movement disturbance	III Discontinuity
Noisiness	Noisiness	—	IV Noisiness
Luminosity	Subopt. Luminosity	—	V Suboptimal Luminosity
—	Degree of degradation	—	—

apart," "torn," and disjointed video. The second dimension *Unclearness* describes a "unclear" or "smeared" image. The third dimension, *Discontinuity*, reflects all interruptions in the flow of the video. The fourth dimension *Noisiness* points to random changes in brightness and color at the level of each pixel. The last and fifth dimension *Suboptimal Luminosity* refers to a too high or low brightness of the video in general. The resulting and fused perceptual video quality space is given in Table 3.13 in column 4.

References

1. Belmudez, B.: AudioVisual Quality Assessment and Prediction for Videotelephony. Springer, GER-Heidelberg (2015)
2. ITU-T Rec. P.930: Principles of a Reference Impairment System for Video. International Telecommunication Union, CH-Geneva, April 1996
3. FFmpeg: A Complete, Cross-Platform Solution to Record, Convert and Stream Audio and Video. https://www.ffmpeg.org. Internet Accessed October 2019
4. MATLAB. version R2017b. Natick, Massachusetts: The MathWorks Inc. (2017)
5. Linux Foundation: Netem—Network Emulation. https://wiki.linuxfoundation.org/networking/netem. Internet Accessed October 2019
6. ITU-T Rec. G.191: Software Tools for Speech and Audio Coding Standardization. International Telecommunication Union, CH-Geneva (2010)
7. Wältermann, M.: Dimension-based Quality Modeling of Transmitted Speech. Springer, GER-Berlin (2013)
8. Côté, N.: Integral and Diagnostic Intrusive Prediction of Speech Quality. T-Labs Series in Telecommunication Services. Springer, GER-Berlin, Heidelberg (2011)
9. ITU-T Rec. G.722.2: Wideband Coding of Speech at around 16 kbit/s using Adaptive Multi-Rate Wideband (AMR-WB). International Telecommunication Union, CH Geneva (2003)
10. TU-T Rec. G.711. Pulse Code Modulation (PCM) of Voice Frequencies. International Telecommunication Union, CH-Geneva (1988)
11. ITU-T Rec. P.910: Subjective Video Quality Assessment Methods for Multimedia Applications. International Telecommunication Union, CH-Geneva, April 2008
12. Kanehara Trading Inc.: Ishihara's Tests for Colour Deficiency—24 Plates Edition (2015)
13. Snellen, H.: Snellen chart. "wikipedia". https://commons.wikimedia.org/wiki/File:Snellen_chart.svg. Accessed August 201

14. Tucker, I.: Perceptual video quality dimensions. Master Thesis, Technische Universität Berlin, GER-Berlin (2011)
15. Möller, S.: Quality Engineering—Qualität kommunikationstechnischer Systeme. Springer, GER-Heidelberg (2010)
16. Guse, D., et al.: TheFragebogen—a standalone questionnaire system that only requires a HTML5-capable web browser. http://www.TheFragebogen.de. Internet Accessed June 2018
17. Schiffner, F., Möller, S.: Audio-Visuelle Qualität: Zum Einfluss des Audiokanals auf die Videoqualitäts- und Gesamtqualitätsbewertung. In: 42. Jahrestag Für Akusitk (DAGA). Deutsche Gesellschaft für Akustik (DEGA e.V.), GER-Aachen (2016)
18. Lebreton, P., Garcia, M.-N., Raake, A.: AVRate: an open source modular audio/visual subjective evaluation test interface (2006). Internet https://github.com/Telecommunication-Telemedia-Assessment/AVRate. Accessed October 10 2019
19. Schiffner, F., Möller, S.: Diving Into Perceptual Space: Quality Relevant Dimensions for Video Telephony. Quality and Usability Lab—Technische Universität Berlin, PO-Lisbon (2016)

Chapter 4
Direct Scaling of Video Quality Dimensions

4.1 Introduction

In the previous chapter, the unknown perceptual video quality space was investigated and identified.

In this chapter, a new method in the domain of video quality assessment is presented, using the five identified quality dimensions. In Table 4.1, these dimensions are listed together with a brief description. This approach is named Direct Scaling (DSCAL) of the perceptual quality space. The term *direct* refers to the fact that each perceptual dimension is rated separately and *directly* by a test participant without any additional mathematical procedure. An indirect example would be, for instance, when a PCA or an MDS is necessary when analyzing results from an SD or PC test. The second advantage of this method is the reduction of experimental effort because the test participant needs to rate fewer rating scales as in e.g. an SD test. Furthermore, more test files or test conditions can be taken into account if one aims for the same test duration. Moreover, gathering more subjective data in an experiment allows quality estimation models, based on perceptual dimensions to be trained more effectively.

The *Direct Scaling (DSCAL)* of the perceptual dimensions allows a feature decomposition of the overall quality. This is realized by relating the overall video quality rating with the ratings from the perceptual dimensions.

The presented method is aimed at non-expert test participants. The ratings scales used for the *DSCAL* method are presented in Sect. 4.2. The test procedure is explained in detail in Sect. 4.3. The *DSCAL* method is verified in three subjective experiments. First, two experiments with *videotelephony* video material are described in Sect. 4.4 and second, one experiment is described for a more general video setting in Sect. 4.5. The chapter closes with a conclusion in Sect. 4.6.

© The Author(s), under exclusive license to Springer Nature Switzerland AG 2021
F. Schiffner, *Dimension-Based Quality Analysis and Prediction for Videotelephony*, T-Labs Series in Telecommunication Services,
https://doi.org/10.1007/978-3-030-56570-1_4

Table 4.1 Description of the perceptual video quality dimensions (VQD) with a brief description

VQD	Name	Description
I	Fragmentation (FRA)	Fallen apart, torn, and disjointed video
II	Unclearness (UCL)	Unclear and smeared image
III	Discontinuity (DIC)	Interruptions in the flow of the video
IV	Noisiness (NOI)	Random change in brightness and color
V	Suboptimal Luminosity (LUM)	Too high or low brightness

4.2 Dimension Rating Scales

The new subjective method provides a means for quantifying the five quality-relevant perceptual dimensions in a passive video consuming setting, by directly rating the five quality descriptive scales. The scales are shown in Fig. 4.1.

Each dimension scale is dedicated to one particular dimension. This enables for directly quantifying separate scores for each perceptual dimension present in the impaired video. The rating scales are continuous 7-point scales. The names of the five dimensions are used as scale titles. The ends of the scale are labeled with antonym pairs to describe the range of the scales. The titles and scale labels in English translation, together with the German terms, are presented in Table 4.2. The continuous

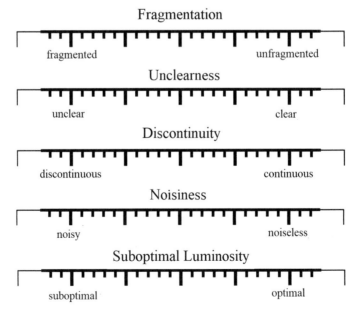

Fig. 4.1 Rating scales for the direct assessment of the quality dimensions in the English translation of the scale titles and labels

Table 4.2 DSCAL scale titles and labels in English translation and original German terms

English scale titles	German scale titles
Fragmentation	Fragmentierung
Unclearness	Undeutlichkeit
Discontinuity	Diskontinuität
Noisiness	Rauschhaftigkeit
Suboptimal Luminosity	Suboptimale Helligkeit
Adjective—ENG (GER)	**Antonym—ENG (GER)**
Fragmented (fragmentiert)	Unfragmented (unfragmentiert)
Unclear (undeutlich)	Clear (deutlich)
Discontinuous (unkontinuierlich)	Continuous (kontinuierlich)
Noisy (verrauscht)	Noiseless (rauschfrei)
Suboptimal (suboptimal)	Optimal (optimal)

nature of the scale possesses the appearance for the test person, that the scales are more sensitive, in comparison to traditional 5-point ACR scales. Nevertheless, it is not proven that this is the case. In the literature, proof is found that the choice of the rating scale is of minor importance [1]. Besides, traditional category scales (like 5-point ACR) tend to have saturation effects at the scale ends, which should be avoided. To attenuate saturation, "overflow" regions are added on each side, similar to the continuous 7-point quality scale used before. The usage of the scales is explained in the exemplary test introduction in Appendix B. On the right side of each scale, the material can be regarded as optimal for that respective property. Thus, the dimension scales can be regarded as unipolar in terms of degradation.

4.3 DSCAL—Test Procedure

This section will provide an overview of the test procedure of the DSCAL method. The method in general follows common paradigms for subjective quality tests [2–4]. The method uses a passive reception approach. The method was verified in three pure video experiments. Each test follows a similar procedure, which is explained in this section. As mentioned in the introduction of this chapter, the method provides means for the quantification of the overall video quality, as well as it is underlying composition. In general, the rating task was divided into two parts, first the overall video quality rating task, and second the dimension rating task. The general rating scheme is shown in Fig. 4.2.

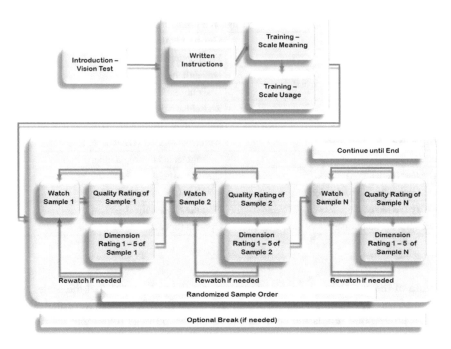

Fig. 4.2 General rating scheme for the DSCAL method used in the three verification experiments (VQDIM I, VQDIM II and VQDIM III)

The participant views the video sample and is allowed to re-watch it as often as necessary, but a least once. The overall video quality rating was done by using the continuous 7-point scale, as described in Sect. 3.2.2. This scale is chosen again since it reduces saturation effects at the scale ends. After the quality rating, the five dimensions are rated separately on the respective scales. Each scale is presented subsequently, but only one at the time in order to reduce the influence the scales could have on each other. Additionally, the order of the dimension scales for each test participant was randomized in order to eliminate the potential influence of order effects. In the beginning, a written introduction was given, and a small training consisting of 6 video samples was performed. This allows the participants to familiarize themselves with the rating tasks and the degradations. The order of the test files was also randomized so that each participant avoids the potential danger of order effects.

Table 4.3 Description of the impairments in the test material

Name	Description
Reference	Unimpaired material
RISV Artificial Blockiness 5 × 5/8 × 8	All frames (2 block sizes settings)
RISV Artificial Blurring ITU(F1)/Filter7	All frames (2 filter settings)
RISV Artificial Jerkiness 6/11 frames	Jerkiness (6 resp. 11 frames hold)
RISV Artificial NoiseQ 3%/15%	Salt & Pepper noise (x% pixel/frame)
H.264 Bitrate 28/56 kbps	H.264-Codec bitrate (2-pass coding)
Packet Loss 0.5%/1.5%	H.264-Codec, traffic control, NetEm
Luminance Impairment I (darker)	Luminance value reduced
Luminance Impairment II (lighter)	Luminance value raised

4.4 DSCAL—Video Telephony

4.4.1 Video Quality Dimension Test I (VQDIM I)

In this section, the first experiment to verify the Direct Scaling (DSCAL) method in the videotelephony context is described.

Test Material: The data set with videos showing short sequences of simulated videotelephony was employed, as explained in Sect. 3.1.1. A subset of video files from the prepared video samples was used. A previous study [5] shows that the test participants can rate the video channel independently from the audio channel. Since the experiment only focuses on the video, the audio channel was muted. The test conditions that are used in this experiment are given in Table 4.3. Examples of the test material are shown in Fig. 3.1.

Test Participants and Procedure: For this study, 15 female and 13 male participants were recruited with an average age of 28.9 years ($\sigma = 7.3$), as summarized in Sect. 3.1.3. No test participant had to be rejected based on abnormal eyesight. The duration of the experiment was approximately 50–60 min.

4.4.1.1 Test Results and Interpretation

The quality scores and the dimension scores from the experiment (VQDIM I) were analyzed. First, the ratings were transformed from the 7-point scale to a 5-point MOS (Mean Opinion Score), as described in [6]. The rating behavior was as expected; the stronger the impairment, the lower the quality rating. The quality scores were in the expected range and compared with the quality scores from the SD-PCA test (see Sect. 3.2.3) and show a high correlation ($r = 0.95$). The RMSE was 0.29 and can be regarded as low. Thus, the quality rating can be regarded as stable (see Table 4.4).

Table 4.4 Quality ratings obtained from two subjective tests (VQDIM I/SD-PCA), scale range from 1–5, whereas 5 represents the highest possible rating and 11 the lowest possible rating

Condition	VQDIM I	SD-PCA
Reference	4.32	4.07
Blockiness 5 × 5	1.77	2.09
Blockiness 8 × 8	1.37	1.52
Blurring ITU (F1)	2.57	2.67
Blurring Filter7	1.74	1.93
Jerkiness 6 frames	2.44	2.94
Jerkiness 11 frames	2.09	2.56
NoiseQ 3%	2.33	2.42
NoiseQ 15%	1.78	1.95
H.264 Bitrate 28 kbps	1.34	1.45
H.264 Bitrate 56 kbps	1.99	2.01
Packet Loss 0.5%	2.99	2.82
Packet Loss 1.5%	2.08	1.74
Luminance Impairment I (darker)	2.72	3.05
Luminance Impairment II (lighter)	2.83	3.08
Pearson R	0.95	
RMSE	0.29	

The dimension scores were averaged over all source materials of the same impairment type, and the statistics mean, standard deviation, and confidence interval for each type of degradation were calculated (see Table 4.5). The results are also depicted in Fig. 4.3. The results mostly showed a very low value for the CI95 with an average of 0.3. In general, the low CI95 means that the rating on the five dimensions is quite consistent. When looking at one level deeper, on the rating of the individual test files, it can be determined that for one type of degradation, the rating is almost identical. This leads to the conclusion that the rating is independent of the source files used. Only one exception can be found in the data set on the *Fragmentation* rating scale. For the test condition *Packet Loss* 0.5%, three of the four source files are rated very similarly (2.11, 2.97, 2.47), and one is rated significantly better (4.35). During the processing of that file, advantageous parts of the video file could be affected, meaning only a B-, or P-frame was affected instead an I-frame. Thus in this test file, the degradation is not that prominent. The error is, therefore, harder to detect or not visible at all. Nevertheless, the test condition *Packet Loss* 0.5% can still be directly linked to *Fragmentation*.

Another finding illustrated in Table 4.5 is that the different degradations clearly trigger the right corresponding dimension (e.g. *Jerkiness 11 Frames* only "hits on the *Discontinuity-Scale*)." Furthermore, one can observe when looking at the dimension ratings that a corresponding impairment in the video had a big negative influence on rating. The rating on the respective dimension scale always sank on average to 3 or

Table 4.5 Dimension scores (VQDIM I—test). Range of the scale 1–5 (Stdev—standard deviation/CI95—95% confidence interval), values below 3 are marked red

Condition	FRA	UCL	DIC	NOI	LUM
Reference	4.57	4.43	4.53	4.54	4.50
Stdev/CI95	0.39 / 0.15	0.53 / 0.20	0.41 / 0.16	0.45/0.17	0.47/0.18
Artificial Blockiness 5 × 5	2.84	2.41	4.26	4.18	4.20
Stdev/CI95	1.30/0.50	1.02/0.40	0.78/0.30	0.93/0.36	0.74/0.29
Artificial Blockiness 8 × 8	2.47	2.11	4.17	4.04	4.07
Stdev/CI95	1.39/0.54	1.07/0.41	0.88/0.34	1.09/0.42	0.90/0.35
Artificial Blurring Filter1	4.42	2.60	4.33	4.38	4.28
Stdev/CI95	0.66/0.26	0.90/0.35	0.68/0.26	0.68/0.26	0.66/0.26
Artificial Blurring Filter7	4.33	1.87	4.31	4.19	4.10
Stdev/CI95	0.74/0.29	0.63/0.24	0.77 / 0.30	0.90/0.35	0.81/0.32
Artificial Jerkiness 6	4.29	3.93	2.27	4.37	4.26
Stdev/CI95	0.72/0.28	0.86/0.33	0.91/0.35	0.67/0.26	0.63/0.25
Artificial Jerkiness 11	4.27	3.88	1.84	4.40	4.26
Stdev/CI95	0.77/0.30	0.85/0.33	0.68/0.26	0.62/0.24	0.65/0.25
Artificial NoiseQ 3%	4.27	3.90	4.36	2.13	4.22
Stdev/CI95	0.92/0.36	0.83/0.32	0.68/0.27	0.66/0.26	0.71/0.27
Artificial NoiseQ 15%	4.22	3.41	4.27	1.57	4.14
Stdev/CI95	0.99/0.38	1.08/0.42	0.82/0.32	0.52/0.24	0.77/0.30
H264 Bitrate 28kbps	2.14	1.97	3.85	3.98	3.95
Stdev/CI95	1.10/0.42	1.04/0.40	0.99/0.39	1.08/0.42	0.94/0.36
H264 Bitrate 56kbps	2.98	2.45	4.06	4.16	4.11
Stdev/CI95	1.05/0.41	0.94/0.36	0.87/0.34	0.83/0.32	0.79/0.31
Packet Loss 0.5%	2.98	3.70	4.09	4.33	4.28
Stdev/CI95	0.81/0.31	0.90/0.35	0.81/0.31	0.63/0.24	0.59/0.23
Packet Loss 1.5%	1.82	3.32	3.88	4.25	4.20
Stdev/CI95	0.63/0.24	1.03/0.40	0.93/0.36	0.74/0.29	0.66/0.26
Luminance Impairment I (darker)	4.43	3.68	4.32	4.43	1.71
Stdev/CI95	0.59/0.23	0.90/0.35	0.73/0.28	0.56/0.22	0.57/0.22
Luminance Impairment II (lighter)	4.50	3.99	4.43	4.45	2.14
Stdev/CI95	0.46/0.18	0.06 / 0.25	0.58 / 0.22	0.50/0.20	0.91/0.35

below. The reason is that all test conditions, besides the *reference*, clearly impair the video and trigger the corresponding video quality dimensions. As a result, an "on-off" rating behavior was obtained. So when the test participants spot a video impairment, they are rating on the intended quality dimension. This leads subsequently to a low rating at the respective dimension scale. Four test conditions are rated on two dimensions (Fragmentation and Unclearness). This was expected, because the blocks in *Blockiness* make the image more fragmented. In the same time, the blocks are blurred through the averaging of the color values. The same holds true for the *Bitrate*

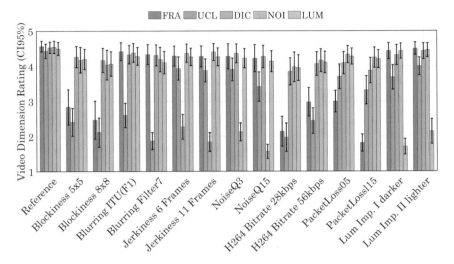

Fig. 4.3 Results of the dimension ratings (VQDIM I)

test condition. Here, the low bitrate produces a smeared image because, with lower bitrate, the high frequencies are cut off. Further, the underlying video codec works are block based, and more and more blocks are visible with a lower bitrate. It is concluded that the scales work as intended. However, a drawback is that the test persons seem to "puzzle." This means they try to find the right category, and if they found the "right" one, it is rated clearly. These results also come from the limited number of test conditions and that one test condition is indented to impair mainly one quality dimension. This was done because otherwise, the test participants would be overexerted with lots of different degrees of impairments. To tackle this limitation, an additional study (VQDIM II) was conducted. This is presented next in Sect. 4.4.2.

4.4.2   Video Quality Dimension Test II (VQDIM II)

In this section, the results for the *VQDIM II* experiment are presented. As briefly discussed in the last section, an "on-off" rating behavior in the rating for the video quality dimensions and partly for the video quality rating was observed. Notwithstanding the fact that valuable information can be derived from the results, it should be investigated how the technical conditions behave in the video quality dimension space.

4.4.2.1 Test Material

To this end, additional test files were prepared (comp. Table 4.6). The aim is to cover the full range of all dimension scales. Furthermore, several combinations of degradations should be investigated. In these cases, each combination of two perceptual quality dimensions should be triggered. This is done to examine if the dimension ratings change when more dimensions are triggered or if the dimension rating stays stable. Would the latter be the case, it would lead to the conclusion that the dimensions are in fact, orthogonal.

Table 4.6 Description of the impairments in the test material for the VQDIM II experiment (single impairments and combination of impairments)

Video impairment—single	Description
Reference	Unimpaired material
RISV Artificial Blurring ITU-Filter	All frames impaired (filter setting 1/3/6 from Rec.)
RISV Artificial Blurring Filter7	All frames impaired (own filter setting)
RISV Artificial Jerkiness X Frames	Jerkiness (3/6/9/12/18 frames held)
RISV Artificial NoiseQ $X\%$	Salt & Pepper noise (1/3/6/9/15% pixel/frame)
H.264 Bitrate xx kbps	H.264-Codec bitrate 2-pass coding (28/56/128/256 kbps)
RISV Artificial Blockiness $A \times A$	Block size (2/5/8/11 pixel)
Packet Loss $x.x\%$	H.264-Codec, traffic control, NetEm 0.3/0.6/1.2/1.8% random packet loss rate
Luminance Impairment I (darker)	Luminance reduced $-25/-50/-75$ (underexposure)
Luminance Impairment II (lighter)	Luminance raised $+25/+50/+75$ (overexposure)
Video impairment—combination	
Blurring + Noise	ITU-Filter 1 + 9% Noise
Blurring + Packet Loss	ITU-Filter 6 + 0.6% Packet loss
Lum-Imp. I + Packet Loss	Lum. reduced -50 + 1.2% Packet Loss
Lum-Imp. I + Blurring	Lum. reduced -50 + ITU-Filter 1
Lum-Imp. II + Noise	Lum. raised $+50$ + 9% Noise
Jerkiness + Blurring	6 Frames + ITU-Filter 1
Jerkiness + Lum-Imp. II	9 Frames + Lum. raised + 50
Jerkiness + Packet Loss	9 Frames + 0.6% Packet Loss
Noise + Jerkiness	9% Noise + 6 Frames
Noise + Packet Loss	9% Noise + 1.2% Packet Loss

4.4.2.2 Test Participants and Procedure

For this study, 19 female and 28 male participants were recruited with an average age of 29.1 years ($\sigma = 4.7$) as summarized in Sect. 3.1.3. No test participant had to be rejected on the basis of abnormal eyesight. The duration of the experiment was roughly 1 h and the procedure is described in Sect. 4.3

4.4.2.3 Test Results and Interpretation

The quality ratings as well as the dimension scores were analyzed. First, the ratings were transformed from the 7-point scale to a 5-point MOS (Mean Opinion Score) scale as before, as described in [6].

Quality Rating: The results from the quality rating are shown in Fig. 4.4 for the single impairments and in Fig. 4.5 for the impairment combinations. The quality ratings show an expected behavior; the stronger the impairment, the lower the quality rating. This holds for each type of single impairments (Blurring, Jerkiness, Bitrate, Packet Loss, etc.). In this experiment, the range of the scale was almost wholly used (MOS_{DIM} 4.26 $-$ 1.26). In addition, the overall video quality ratings, from conditions with two impairments are lower, when compared to the single impairments of the same type in general. This result was also expected since each degradation itself reduces the rating. This underlines the additive nature of degradation on quality.

Video Dimension Rating (Single Impairments): In this paragraph, the ratings for the different types of degradation in the video material are investigated. The focus here lies on the examination, whether the impairments are capable of triggering the

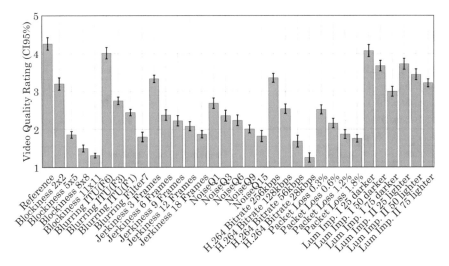

Fig. 4.4 Results (MOS) of the quality ratings for the single impairments (VQDIM II)

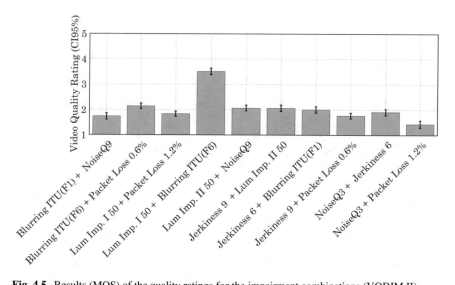

Fig. 4.5 Results (MOS) of the quality ratings for the impairment combinations (VQDIM II)

"right" corresponding quality dimensions. In addition, the different degrees of one type of impairment are indented to be distributed over the whole range of the MOS scale. This is meant to unveil how the different types of impairments behave in the perceptual video quality space.

Fragmentation: The results are shown in Fig. 4.6. Regarding the dimension ratings for *Fragmentation*, the test conditions *Jerkiness, Noise, Luminosity Impairment I & II, and Blurriness* had no negative impact. The rating almost stays the same with a MOS clearly above 4. The test conditions *Blockiness, Packet Loss*, and *Bitrate* had a clear negative impact on the rating. This was expected since all test conditions are impairing the unity of the image. *Blockiness* had the smallest negative impact, but still reduces the rating below 3. The bigger the blocks, the more fragmented the video image looks. The same is true when reducing the bitrate. This leads to more and more visible block-like artifacts in the video imagery. The test condition *Bitrate* yields a linear rating behavior with a spread to the ratings over the whole range of the scale. *Packet Loss* had the strongest negative impact on the ratings, where even relatively low packet loss rates dropped the rating to an MOS of 2 and below. When regarding *Packet Loss*, it is obvious that "slicing-," "partly green-out-" artifacts typical for that type of impairment are perceived as fragmented and had, therefore, a low rating.

Summarizing, the rating behavior on that scale was as expected and shows that the scale works as indented.

Unclearness: The test conditions *Luminosity Impairment I & II* and *Jerkiness* show no significant differences in the ratings. For the test conditions *Noise*, there is a slight influence on the *Unclearness* rating. This can be explained because noise is added to the video image, and the image becomes more and more superimposed, leading

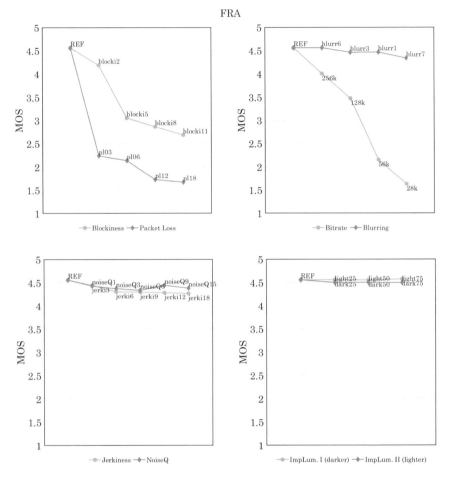

Fig. 4.6 Results (MOS) for the perceptual video quality dimension *Fragmentation* (FRA) for all single impairments. X-axis ranging from Reference (REF)—no impairment to higher degrees of impairments.

to unclear imagery. This effect is undeniable but still only of low impact. Moreover, the impact is nearly identical for all degrees of noise added to the video.

The test conditions that are indented to have a strong negative impact on the *Unclearness* rating are *Blockiness*, *Blurring*, and *Bitrate*. All ratings for the three impairment types are distributed over the entire MOS scale. The *Blurring* impairment is directly aimed to make the video unclear, so a negative influence is not surprising. Also, the *Unclearness* rating gets reduced the lower the bitrate. This is due to the fact that when reducing the bitrate, the higher frequencies become cut off more and more leading to a smeared imagery. When looking at *Blockiness*, it is clear that when the blocks are getting bigger, the ratings drop. Since in the creation of that impairment, the color and luminance values in one block are averaged, the details in

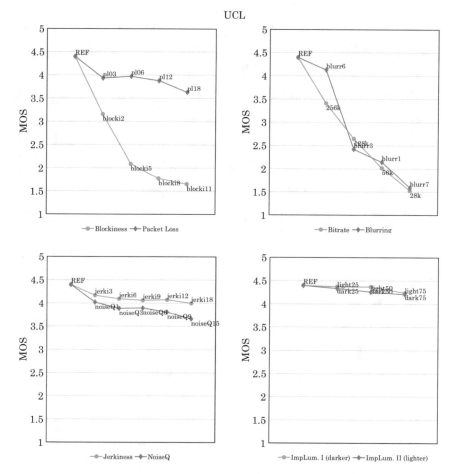

Fig. 4.7 Ratings (MOS) for the perceptual video quality dimension *Unclearness* (UCL) for all single impairments. X-axis ranging from Reference (REF)—no impairment to higher degrees of impairments

color, contrast, etc., are lost. This leads to an unclear image in each block, and that is precisely what is represented in these ratings. The results are shown in Fig. 4.7.

Discontinuity: The ratings for the video quality dimension *Discontinuity* are shown in Fig. 4.8. On the first view, one can see that, as indented, the test condition *Jerkiness* clearly has a negative impact on this respected dimension scale. In addition, the two test conditions *Packet Loss* and *Bitrate*. Since the artifacts of packet loss interrupt the flow of the video, it is rated slightly as discontinuity, even when only parts of one video frame are affected. Considering *Packet Loss*, this slight negative impact on the rating shows almost the same behavior on the whole scale.

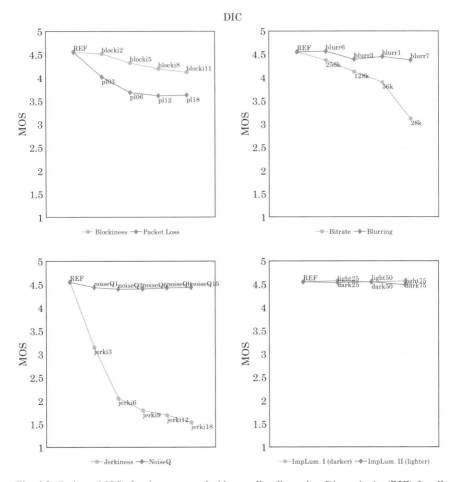

Fig. 4.8 Ratings (MOS) for the perceptual video quality dimension *Discontinuity* (DIC) for all single impairments. X-axis ranging from Reference (REF)—no impairment to higher degrees of impairments.

The test condition *Bitrate* 28 kbps also had a negative impact on the rating, but not as severe as *Jerkiness*. When viewing the files with the very low bitrate, it stands out that the video image is built stepwise. The decoding process uses previous frames to construct the video image. This means that the appearance of the video looks different in terms of quality from frame to frame. These changes lead to a perceived discontinuity. All results for the *Discontinuity* scale are plausible, and it works as indented.

Noisiness: When investigating the ratings on the *Noisiness* scale, it becomes clear the only the *Salt & Pepper* noise conditions had an impact here. An interesting effect is that the rating drops roughly to an MOS of 2, even when there is only a mild

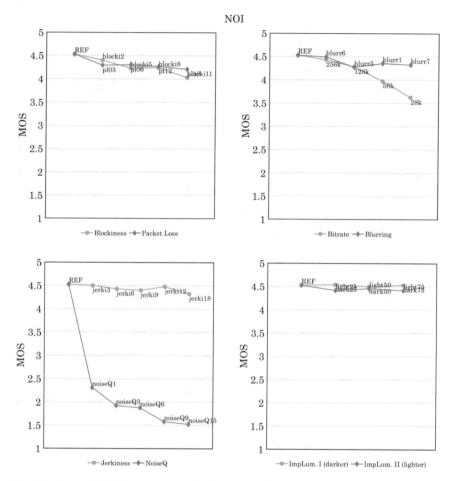

Fig. 4.9 Ratings (MOS) for the perceptual video quality dimension *Noisiness* (NOI) for all single impairments. X-axis ranging from Reference (REF)—no impairment to higher degrees of impairments

impairment present (only 1% of the pixel in each frame are noisy). The only exception are the conditions *Bitrate* 28 and 56 kbps. The coding artifacts are perceived slightly as noisy when the bitrate is very low.

Nevertheless, the scale works as intended, and the resulting ratings are as expected and are shown in Fig. 4.9.

Suboptimal Luminosity: When looking at the ratings on the *Suboptimal Luminosity* scale in Fig. 4.10, one can see that only the conditions with over- and underexposure (comp. bottom left in Fig. 4.10) had a negative impact on the dimension rating. Nearly the whole scale was used linearly. There is only one exception in the test condition set, namely, *Bitrate* 28 kbps. Due to the low bitrate, the contrast in the videos is very

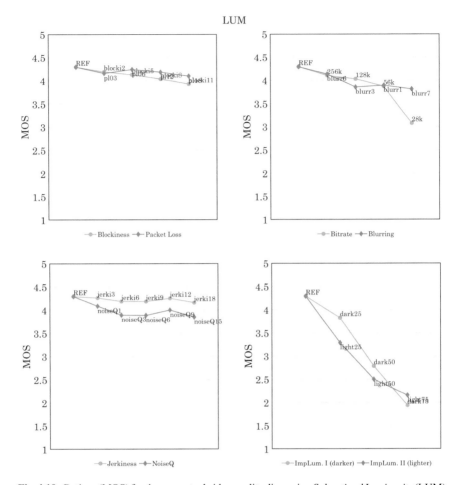

Fig. 4.10 Ratings (MOS) for the perceptual video quality dimension *Suboptimal Luminosity* (LUM) for all single impairments. X-axis ranging from Reference (REF)—no impairment to higher degrees of impairments

low, and the whole video looks washed out. Since that test condition represents a severe degradation in the video, The rating drops to around 3.

In summary, it can be said that the scale for *Suboptimal Luminosity* works as intended.

Video Dimension Rating (impairments Combination): In this paragraph, the ratings for the different combinations of degradations in the video material are investigated. The focus here lies on the examination, whether the ratings for one particular type of impairment stays the same even when there is an additional impairment aimed to trigger another perceptual dimension. The results are shown in Fig. 4.11. In all five figures, the green line-graph is the baseline of the single impairment. The combined impairments are placed in the same categories as the accompanying single impairments. One can observe, that even when a second impairment is present in the video, the ratings on the respective scale is almost identical with no significant differences. The only exception is that the test condition *noiseQ9* superimposes slightly the negative impact of the test condition *light50* and *blurr1*. In both cases, the rating is roughly 0.5 MOS better compared to the single impairment. In the too bright and noisy test condition, the noise added darkens the video to some degrees, and therefore the rating is better than the too bright only test condition. In the blurred and noisy test condition, the conjecture is that the noise adds more information back to the image and with that masks the smeared effects.

In summary, all five perceptual video quality dimension scales work as intended. The scales are sensitive enough to detect negative influences of the impairments with regards to the corresponding quality dimensions. In all cases, when an impairment type had a negative influence on a quality dimension, the whole range of the scale was used. The ratings also spread along this axis for different degrees of the respective impairment. The main goal of this experiment was successfully obtained, and each dimension was measured. In addition, it could be proved that the ratings for different impairments did not change when a second impairment, triggering a second quality dimension, was added. This reveals that the quality dimensions for video can independently be measured from each other.

4.5 DSCAL—Other Video Contents

4.5.1 Video Quality Dimension Test III

The perceptual video quality space was mostly investigated within the context of videotelephony, and the resulting question arises: Is the perceptual video quality space also valid in a broader, more general video setting? This is meant to also underpin the concept that the perceptual quality space for transmitted video is universal and independent from the content. Furthermore, another goal was to investigate whether the DSCAL method is also applicable to other video contents.

Test Material The main criteria for choosing the test material was that it needed a wide range of content representing today's video- and video streaming footage. In addition to content mostly covered by traditional media, it was essential to consider relatively new media platforms like *YouTube* and *Vimeo*. With a large proportion of user-generated content and their ever-increasing role in the overall media landscape, content that is characteristic for these platforms had to be considered for the test [7]. This also included self-recorded video game footage and online-multiplayer-games enjoyed by a large number of online followers. Due to the enormous impact that video games have had over the past few decades on society and culture [8], it was decided to include footage from two different video games that represent two typical gaming classes.

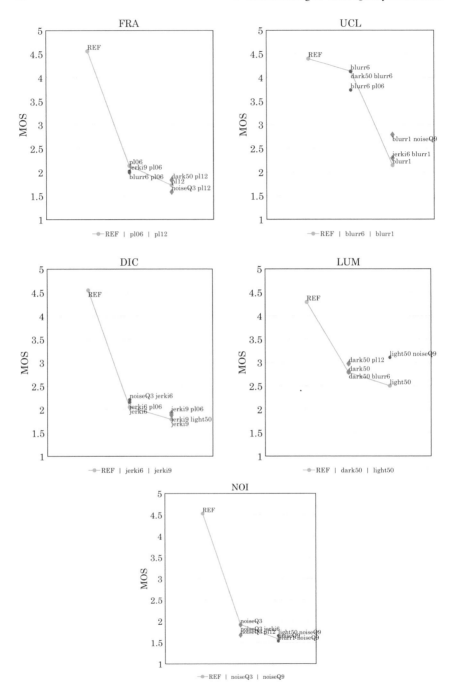

Fig. 4.11 Ratings (MOS) for the perceptual video quality dimensions for all impairment combinations. X-axis ranging from Reference (REF)—no impairment to higher degrees of impairments. The green graph shows the single impairment as a baseline, and besides, the impairment combinations are shown

Table 4.7 Description of the impairments in the test material

	Content class	Description
$C1 - V1$	Movie	Woman walking (steady-cam)[a]
$C1 - V2$	Movie	Boat slowly moving towards the camera[a]
$C2 - V3$	Synthetic	Animation film *Big Buck Bunny*[b]
$C2 - V4$	Synthetic	First-person shooter video game *Crysis*[c]
$C2 - V5$	Synthetic	Online multiplayer video game *Dota 2*[d]
$C3 - V6$	IPTV	Women facing camera (VLOG)[a]
$C3 - V7$	IPTV	Recorded college football match[e]

[a]Produced by Vladimir Bondarenko
[b]All rights by Blender Foundation; https://peach.blender.org/download/
[c]All rights by Electronic Arts; https://www.ea.com/de-de/games/crysis/crysis
[d]All rights by Value Corporation; http://de.dota2.com
[e]Taken from http://vimeo.com/, creative commons licence CC by 3.0

Fig. 4.12 Content classes: C1 (Movie), C2 (Synthetic) and C3 (IPTV)

Therefore, three *Content Classes* were defined, namely, *Movie*, *Synthetic*, and *IPTV* (see Table 4.7 and Fig. 4.12). The footage for the experiment was partly taken from *vimeo.com* under the creative commons license CC-BY 3.0 [9] and partly self-produced by members of the Quality and Usability Lab.

The experiment only focuses on the video; therefore, the audio channel was muted. The aim was to cover a broad range of potential video degradations appearing in video

Table 4.8 Overview—properties of processed test material

Material	7 different source videos
File durations	Approx. 10 s
No. of test conditons	10 (+1 reference)
No. of files to rate	77 (+6 training)
Audio channel	Muted
Resolution	1920 × 1080 pixel (full HD)
Frame rates	24–30 fps
Bit depth	10 bit
Chroma subsampling	4 : 2 : 2
Container	(.mp4)
Viewing distance	ca. 60 cm

Table 4.9 Description of the impairments in the test material

Name	Description
Reference	Unimpaired material
RISV Artificial Blurring ITU (F1)	Filter setup no. 1
RISV Artificial Blurring Filter7	Own adjusted filter
RISV Artificial Jerkiness 3 frames	Jerkiness (3 frames holded)
RISV Artificial Jerkiness 6 frames	Jerkiness (6 frames holded)
RISV Artificial NoiseQ 3%	Salt & Pepper (3% pel/frame)
RISV Artificial NoiseQ 15%	Salt & Pepper (15% pel/frame)
Packet Loss 0.25%	H.264, traffic control, NetEm
Packet Loss 0.5%	H.264, traffic control, NetEm
Luminance Impairment I (darker)	Luminance value (reduced)
Luminance Impairment II (lighter)	Luminance value (raised)

transmission, similar to the experiments presented before. For more details on the test material in this experiment see Tables 4.8 and 4.9. Examples of the degraded test material are shown in Fig. 4.13.

Test Participants As in the previous experiments, the recruited 25 test participants were tested for normal eyesight (Ishihara test—to measure color deficiencies [10], Snellen table—to measure visual acuity [11]) (comp. Sect. 3.1.3). Some participants did not meet the requirements on the Snellen chart but were nonetheless permitted to participate in the experiment. Their results were marked, and it was necessary to examine whether their rating behavior significantly deviated from the rest of the group. In this case, these data sets would be excluded from the analysis. The test procedure was as described in Sect. 4.3.

Fig. 4.13 Examples for processed footage (from left to right, top to bottom): Blurring $F1$, Blurring $Filter7$, Jerkiness 3 (frames), Jerkiness 6 (frames), Luminance (darker), Luminance (lighter), Noise 3%, Noise 15%, Packet Loss 0.25%, Packet Loss 0.50%

4.5.1.1   Analysis of Potential Outliers

Some test participants had difficulties with the visual acuity test. It was not expected to have a significant impact on the assessment, but an analysis is still required on whether some participants show anomalies in their rating behavior. Since the mean or the standard deviation as methods for detecting outliers are quite sensitive [12], the method of (MAD) was applied in the process of finding outliers.

In interpreting the results of the MAD analysis, a very conservative approach was established as a threshold.[1] Based on the analysis, it can be argued that the test participants with impaired vision did not show substantially abnormal behavior in the assessment of video quality and video dimensions. The analysis also unveiled that participants who deviate from the mean tend to display similar deviations across all dimensions and quality ratings. This means that if a person decided to give a rating closer to the edges of the rating scale, the proclivity for extremes would be seen throughout all the assessments. Likewise, if a person displays an aversion to rate in the extremes, this behavior is also reflected in all areas. It exceeds the scope of this work and the available data to discuss the underlying dynamic of each person. In conclusion, the data produced by the participants with weaker visual acuity was of the same goodness as from the participants with normal eyesight. Their data was therefore included in further analysis.

4.5.1.2   Data Analysis

In the following, the quality scores and the dimension scores obtained from the experiment were analyzed (see Table 4.10).

Quality Scores The quality ratings showed an expected behavior. The stronger the impairment, the lower the quality rating becomes. The ratings were compared with the quality scores from the VQDIM I test (see Sect. 4.4.1). This showed a strong linear correlation ($r = 0.71$) between the two independent subjective tests. Thus, the quality rating can be regarded as stable for the different test degradations, even if the video content is diverse. Regarding only the quality scores reveals that the higher the impairment in the video, the worse the quality rating, as expected. Packet Loss has the strongest negative influence on the overall video quality. It can be argued that this is a result of PL triggering two perceptual dimensions (see next paragraph), and as a result has a more significant influence on the overall video quality. Also, a large amount of noise in the video, as well as a highly smeared video, have a strong negative impact on the quality. The luminance degradations do not have the same influence for under- and overexposure on the quality as expected. Underexposure was rated much better than overexposure. The reason for that could be that the static raise/reduction to a fixed value was unfortunately chosen wrong. The perceptual effects seem to be nonlinear. As it is broadly understood, the psychometric function

---

[1]Miller   [13]   proposed   MAD   threshold   criteria: $very\,conservative = 3$, $moderately\,conservative = 2.5$, and $poorly\,conservative = 2$ [13].

of the human perception is mainly S-shaped with a roughly linear part around the 50% perception threshold. The choice to raise/reduce the luminance value by the same value seems to exceed the approximately linear part and reached into the saturation part of the psychometric function. It would be interesting to investigate under- and overexposure in this context in more detail in future work.

Dimension Scores The dimension scores were averaged over all source materials of the same impairment type. Table 4.10 illustrates that the different degradations clearly trigger the right corresponding dimensions (e.g.: *Jerkiness* only "hits" on *Discontinuity*). Only the two PL conditions showed low values on two dimension scales, namely, *Fragmentation* and *Discontinuity*. It can be argued that during the treatment of the files with NetEm and FFmpeg, an unusually high number of frames were dropped due to packet loss. This results subsequently in a stuttering playback in addition to other packet loss typical artifacts. These aspects when combined are the cause supposed for triggering the two dimensions. Nonetheless, the *Fragmentation* rating was almost one MOS worse than the *Discontinuity* rating.

Comparing the dimension ratings from the VQDIM I study in Sect. 4.4.1 with the dimension ratings from this study, it shows virtually no difference in the rating behavior. A *Pearson correlation* with a significance level of 0.05 was calculated (see Table 4.11) to test the degree to which there is a positive relationship between the ratings. With an average mean of $r = 0.82$ and all p-values falling under 0.05, it can be stated that there is a significant positive relationship between the ratings of the two separately conducted tests. It can be assumed that the *multidimensional perceptual video quality space* is stable across video contents. This underlines the assertion that the perceptual video quality space is independent of a wide variety of video content.

Analysis of Content Classes and the Individual Videos In this section, the rating behavior between the different content classes ($C1 - C3$) and the videos ($V1-V7$) will be investigated. The analyzed results are shown in Table 4.12.

As it can be observed, each column that represents a dimension (e.g. *Fragmentation, Unclearness*, etc.) has a confidence interval σ that lies between 0.08 and 0.13. These results already point to a very homogeneous rating behavior, which would suggest that it is content-independent. Further, a Mauchly's Test of Sphericity was calculated, which indicated that the assumption of sphericity had been violated ($X^2(20) = 41.293, p = 0.004$). Therefore, an ANOVA with repeated measures with Greenhouse–Geisser correction was calculated ($F(3.757, 90.161) = 2.021, p < 0.102$). This means that there are no significant differences between the ratings of the different videos. The independence becomes even more apparent when the bar graphs in Figure 4.14 are carefully observed. When the video ratings from Table 4.12 are placed into the corresponding categories, the differences between the individual bars are barely visible. In addition to the visible similarities between the groups, the relatively low-quality ratings stand out compared with the other bar graphs. This could be because each condition has a potentially perceivable impact on the overall quality of the file, whereas the individual dimensions are only triggered when the corresponding conditions appear. Therefore, the quality is more often negatively affected by conditions than the dimensions.

Table 4.10 Results: Quality and Dimension scores. Range of the scale 1 (worst)—5 (best), Q—Quality, FRA—Fragmentation, UCL—Unclearness, DIC—Discontinuity, NOI—Noisiness, LUM—Suboptimal luminosity, the two lowest values for each quality dimension are marked red, (Stdev—standard deviation/CI95—95% confidence interval)

Condition	Q	FRA	UCL	DIC	NOI	LUM
Reference	4.47	4.53	4.45	4.42	4.56	4.40
Stdev/CI95	0.49/0.07	0.45/0.06	0.49/0.07	0.56/0.08	0.32/0.04	0.51/0.07
Artificial Blurring Filter1	3.01	4.17	2.59	4.09	4.12	3.72
Stdev/CI95	0.75/0.11	0.81/0.12	0.94/0.14	0.87/0.13	0.86/0.13	0.90/0.13
Artificial Blurring Filter7	2.24	4.10	1.95	3.89	3.77	3.53
Stdev/CI95	0.69/0.10	0.88/0.12	0.77/0.11	1.03/0.14	1.12/0.16	0.93/0.13
Artificial Jerkiness 3	3.40	4.23	4.03	2.92	4.35	4.10
Stdev/CI95	0.85/0.13	0.80/0.12	0.77/0.11	1.15/0.17	0.56/0.08	0.73/0.11
Artificial Jerkiness 6	2.90	4.07	3.92	2.13	4.27	4.08
Stdev/CI95	0.88/0.12	0.96/0.13	0.84/0.12	0.94/0.13	0.69/0.10	0.70/0.10
Artificial NoiseQ 3%	2.68	4.16	3.43	4.10	2.01	3.69
Stdev/CI95	0.75/0.11	0.82/0.12	1.03/0.15	0.88/0.13	0.70/0.10	0.89/0.13
Artificial NoiseQ 15%	2.26	3.89	3.07	4.07	1.66	3.50
Stdev/CI95	0.79/0.11	1.10/0.15	1.07/0.15	0.87/0.12	0.53/0.07	0.97/0.13
Packet Loss 0.25%	2.05	1.72	3.25	2.49	3.95	3.88
Stdev/CI95	0.81/0.12	0.61/0.09	1.05/0.16	1.15/0.17	0.99/0.15	0.82/0.12
Packet Loss 0.5%	2.00	1.67	3.14	2.53	3.81	3.88
Stdev/CI95	0.79/0.11	0.50/0.07	1.03/0.14	1.11/0.15	1.09/0.15	0.79/0.11
Lum. Impair.I (darker)	3.95	4.49	4.30	4.39	4.48	3.14
Stdev/CI95	0.65/0.09	0.47/0.07	0.59/0.08	0.62/0.09	0.44/0.06	1.24/0.17
Lum. Impair.II (lighter)	3.06	4.38	3.80	4.30	4.30	2.07
Stdev/CI95	0.75/0.11	0.56/0.08	0.90/0.13	0.68/0.10	0.62/0.09	0.85/0.13

Table 4.11 Correlation between the dimension scores from the VQDIM III test with the dimension scores from the VQDIM I test

	FRA	UCL	DIC	NOI	LUM
Corr. (Pearson; CI = 95%)	0.73	0.92	0.85	0.98	0.80
p-value	0.00	0.00	0.00	0.01	0.00

Table 4.12 Quality and dimension scores of the different content classes and per video. Range of the scale 1 (worst)—5 (best), Q—Quality, FRA—Fragmentation, UCL—Unclearness, DIC—Discontinuity, NOI—Noisiness, LUM—Suboptimal luminosity, (Stdev—standard deviation/CI95—95% confidence interval), σ—standard deviation for the rating on each scale

Content class	Q	FRA	UCL	DIC	NOI	LUM
C1	2.88	3.79	3.48	3.53	3.68	3.57
Stdev/CI95	1.06/0.09	1.24/0.10	1.12/0.09	1.27/0.10	1.27/0.10	1.06/0.09
C2	2.95	3.81	3.48	3.61	3.80	3.68
Stdev/CI95	1.08/0.07	1.21/0.08	1.15/0.08	1.21/0.08	1.20/0.08	1.04/0.07
C3	2.87	3.68	3.38	3.57	3.75	3.66
Stdev/CI95	1.04/0.08	1.30/0.10	1.14/0.09	1.24/0.10	1.18/0.09	1.03/0.08
Video						
V1	2.82	3.77	3.42	3.50	3.55	3.51
Stdev/CI95	1.02/0.12	1.22/0.14	1.12/0.13	1.24/0.14	1.32/0.15	1.04/0.12
V2	2.95	3.82	3.54	3.56	3.81	3.64
Stdev/CI95	1.10/0.13	1.26/0.15	1.11/0.13	1.30/0.15	1.20/0.14	1.08/0.13
V3	3.07	3.90	3.64	3.78	3.88	3.82
Stdev/CI95	1.11/0.13	1.17/0.13	1.13/0.13	1.20/0.14	1.19/0.13	1.05/0.12
V4	2.81	3.74	3.23	3.44	3.71	3.46
Stdev/CI95	1.04/0.12	1.22/0.14	1.13/0.13	1.18/0.13	1.19/0.13	1.06/0.12
V5	2.97	3.80	3.56	3.61	3.82	3.76
Stdev/CI95	1.09/0.12	1.23/0.14	1.16/0.13	1.25/0.14	1.20/0.14	0.99/0.11
V6	2.92	3.78	3.48	3.67	3.83	3.66
Stdev/CI95	1.08/0.12	1.25/0.14	1.16/0.13	1.22/0.14	1.17/0.13	1.04/0.12
V7	2.82	3.57	3.28	3.47	3.67	3.66
Stdev/CI95	0.99/0.11	1.34/0.15	1.11/0.13	1.26/0.14	1.18/0.13	1.03/0.12
σ	0.08	0.09	0.13	0.10	0.10	0.11

4.5.1.3   Conclusion

This experiment shows that the perceptual quality space for video can be assessed directly via the corresponding quality dimensions for different video contents. The method of Direct Scaling (DSCAL) was also, in this case, easy to use, and one can obtain meaningful dimension scores. The presence of a wide variety of video content does not affect the perceptual video quality space. Therefore, the perceptual video quality space is valid for video in general and not only in the video telephony domain. This substantiates the idea of universal perceptual quality space for video.

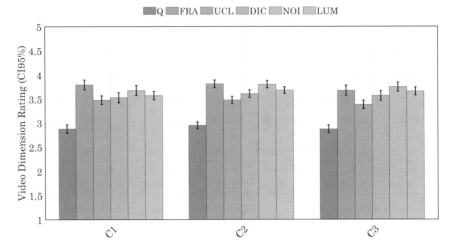

Fig. 4.14 Quality and dimensional ratings divided into the 3 categories Movie (C1), Synthetic (C2), and IPTV (C3)

4.6 DSCAL—Conclusion

In this chapter, the methodology of Direct Scaling (DSCAL) was presented. This methodology introduced the rating of the perceptual video quality dimensions in a direct way. With direct is meant in this context that no additional mathematical procedure (like PCA & MDS) for gathering ratings of underlying perceptual dimensions is needed. It allows naïve test subjects to rate each perceptual dimension separately and directly quantify them. Therefore, a test stimulus that is degraded in one or several dimensions can be rated, and its composition regarding the perceptual dimension can directly be observed. In the three experiments described in previous sections, it could be shown that non-expert test subjects can distinguish between the five perceptual video quality dimensions and rate them accordingly. In this method, no long introduction phase is required to train the test subjects for the rating task.

By comparing the results of the three studies (VQDIM I–III), evidence was provided that the application of the DSCAL method leads to comparable and useful results. Especially in regards to the experiment VQDIM II (see Sect. 4.4.2). This experiment showed that the dimension scales measure what they are constructed for. Moreover, the respective dimension ratings of conditions impaired in two perceptual dimensions correspond to those of the corresponding conditions impaired in one dimension. No relevant interaction could be observed so far.

To substantiate that the PEARSON correlation between the dimension scales were calculated. The results from the VQDIM II experiment (Sect. 4.4.2) were used here. The resulting correlation coefficients are presented in Table 4.13 and show that there is besides one case, no linear relationship. Only the correlation between FRA and UCL has a weak positive relationship, but the value is minimal and can be neglected.

Table 4.13 Correlation coefficients from the five-dimensional scales, calculated from the VQDIM II experiment

	FRA	UCL	DIC	NOI	LUM
FRA	1	0.33	0.22	−0.05	−0.22
UCL		1	−0.06	0.02	−0.16
DIC			1	−0.11	−0.23
NOI				1	−0.03
LUM					1

To summarize: the method can be regarded as a *tool* to analytically assess transmitted video in order to diagnose the nature of the overall quality and its build-up elements.

The purpose of DSCAL was to create a method to rate the perceptual video quality space in a direct way. In addition, the rating process should be simplified in terms of time savings and effort reduction. All three targets have been reached. To be specific, this method produces meaningful data and reduces the experimental effort since the test participants only need to rate 6 scales instead of e.g.18 scales in the SD-PCA test. Furthermore, more test files or test conditions can be taken into account if one aims for the same test duration. In the VQDIM I test twice as many test files could be investigated compared to the SD-PCA test.

The introduced method was made available to the ITU-T Study Group 12 [14]. Additional contributions in this realm were made, and based on those, Study Group 12—Question 7 implemented a work item called P. VQD. The work led to the implementation of the new ITU-T Recommendation P.918, named *Dimension-based Subjective Quality Evaluation for Video Content*.

References

1. Tominaga, T., Hayashi, T., Okamoto, J., Takahashi, A.: Performance comparisons of subjective quality assessment methods for mobile video. In: 2010 Second International Workshop on Quality of Multimedia Experience (QoMEX). IEEE (2010)
2. ITU-T Rec. P.800: Methods for subjective determination of transmission quality. International Telecommunication Union, CH-Geneva (1996)
3. ITU-T Rec. P.910: Subjective video quality assessment methods for multimedia applications. International Telecommunication Union, CH-Geneva, April 2008
4. ITU-T Rec. P.911: Subjective audiovisual quality assessment methods for multimedia applications. International Telecommunication Union, CH-Geneva, December 1998
5. Schiffner, F., Möller, S.: Audio-Visuelle Qualität: Zum Einfluss des Audiokanals auf die Videoqualitäts- und Gesamtqualitätsbewertung. In: 42. Jahrestag Für Akusitk (DAGA). GER-Aachen: Deutsche Gesellschaft für Akustik (DEGA e.V.) (2016)
6. Köster, F., et al.: Comparison Between the Discrete ACR Scale and an Extended Continuous Scale for the Quality Assessment of Transmitted Speech. DAGA, GER-Nürnberg (2015)

7. YouTube Inc.: Over one billion users—that's how many users YouTube has—meaning almost one-third of the Internet. https://www.youtube.com/yt/about/press/. Internet: Accessed June 2018

8. Hasan, M.R., Jha, A.K., Liu, Y.: Excessive use of online video streaming services: impact of recommender system use, psychological factors, and motives. Comput. Hum. Behav. **80**, 220–228 (2018)

9. Creative Commons. Attribution 3.0—CC BY 3.0. https://creativecommons.org/licenses/by/3.0/. Internet Accessed June 2018

10. Kanehara Trading Inc.: Ishihara's Tests for Colour Deficiency—24 Plates Edition (2015)

11. Snellen, H.: Snellen chart. "wikipedia". https://commons.wikimedia.org/wiki/File:Snellen_chart.svg. Accessed August 2017

12. Leys, C., et al.: Detecting outliers: do not use standard deviation around the mean, use absolute deviation around the median. J. Experimental Soc. Psychol. **49**(4), 764–766 (2013)

13. Miller, J.: Short report: reaction time analysis with outlier exclusion: bias varies with sample size. In: The Quarterly Journal of Experimental Psychology Section A (1991)

14. Schiffner, F., Möller, S.: Proposal for a draft new recommendation on dimensions based subjective quality evaluation for video content. ITU-T Study Group 12—Contribution C.291. International Telecommunication Union—Telecommunication Standardization Sector, CH-Geneva (2018)

Chapter 5
Quality Modeling and Prediction

5.1 Dimension-Based Modeling of Overall Video Quality

This chapter presents a linear approach to predict the overall quality for video in terms of Mean Opinion Score (MOS) from obtained dimensional ratings. Furthermore, it presents a comparison of the model with the quality ratings of independent subjective experiments.

5.1.1 Linear Quality Model

To analyze the relationship between the dimension ratings of the five underlying perceptual dimensions (Fragmentation (FRA), Unclearness (UCL), Discontinuity (DIC), Noisiness (NOI), Suboptimal Luminosity (LUM)) and the overall video quality, a model using linear regression was calculated. The data from the VQDIM I—test (see Sect. 4.4.1) was used to train the model. The model is calculated on the averaged dimension ratings on the level of test condition. The dependent variable was the quality rating, and the independent variables are the five perceptual dimensions.

The formula for the linear model is as follows:

$$MOS_{pre-lin} = \\ -5.94 + (0.21 \cdot FRA) \\ +(0.73 \cdot UCL) + (0.53 \cdot DIC) \\ +(0.47 \cdot NOI) + (0.31 \cdot LUM)$$

The model shows that the dimension Unclearness (UCL) has the most significant influence and followed by Discontinuity (DIC). The next dimensions in order are Noisiness (NOI) and Suboptimal Luminosity (LUM). The dimension Fragmentation (FRA) has the lowest influence in the model. This finding was unexpected, but in the

© The Author(s), under exclusive license to Springer Nature Switzerland AG 2021
F. Schiffner, *Dimension-Based Quality Analysis and Prediction for Videotelephony*, T-Labs Series in Telecommunication Services,
https://doi.org/10.1007/978-3-030-56570-1_5

Packet Loss—test conditions blurring can also be found to a certain degree. These test conditions trigger FRA but are also pointing to UCL. Also, the test conditions *H.264 bitrate* and *Blockiness* trigger both dimensions (FRA & UCL) and are perceived as fragmented and unclear.

The calculated ANOVA reports a significant F value ($F(5, 9) = 53.38$, $p < 0.001$), indicating that the model works well. In general, the linear regression does an excellent job of modeling the quality ratings from the dimension scores. The coefficient of determination is $r^2 = 0.967$ and, therefore, can be regarded as excellent. The model explains nearly 97% of the variation in ratings and has a standard error of the estimation of 0.17. These high values could also be influenced by the fact that the dimensions are presented in a group, even if they were shown on separate screens.

5.1.2   Quality Modeling—VQDIM I

After the model was calculated, the dimension scores, obtained from the VQDIM I—test, are sent to the model. The predicted MOS ($VQDIM\ I_{esti}$) were compared to transformed quality ratings from two subjective experiments (VQDIM I, SD-PCA). The model predicts the quality ratings well, as can be seen in Table 5.1 (column 2–3) and Fig. 5.1. The correlation of the subjective test results with predicted MOS ($VQDIM\ I_{esti}$) is high in both cases ($r_{VQDIM\ I} = 0.98$, $r_{SD-PCA} = 0.94$). The RMSE is very low for $VQDIM\ I_{esti}\ versus\ VQDIM\ I$ (RMSE $= 0.13$) and still acceptably low for $VQDIM\ I_{esti}\ versus\ SD - PCA$ (RMSE $= 0.39$).

This might have been expected since, in both experiments, the source material and the degradations were the same. However, the test participants were different in both experiments, and the lab setting differed slightly.

Moreover, a second linear quality model (SD model) was calculated for comparison using all ratings from Antonym Pairs of the SD test presented in Sect. 3.2.3.

These predicted MOS values are also shown in Table 5.1 (see column 4 *SD Model*). These values were correlated with the predicted MOS values ($VQDIM\ I_{esti}$). The correlation is very strong and the coefficient is $r_{VQDIM\ I_{esti}}\ versus\ SD model = 0.93$. In Fig. 5.1, this result is depicted, and one can observe the strong linear relationship between the two models. Since the correlation between the two linear quality prediction models is very high, it is argued that the introduced method of DSCAL should be used. As said, this method is easier, and one also obtains meaningful dimension scores for the later overall quality modeling.

5.1.3   Quality Modeling—VQDIM II

As described, the good result of the model is partly explained by the use of the same data for model training and testing. In this section, the linear model is investigated in terms of its performance, when using data from other studies. Therefore, the

Table 5.1 Comparison of the transformed quality ratings from the two subjective experiments (*VQDIM I, SD-PCA*) with the predicted MOS from the two linear models (*V QDIM I$_{esti}$* and Semantic Differential (SD)-model)

Condition	MOS VQDIM I	MOS SD-PCA	$VQDIM\ I_{esti}$ Model	SD Model
Reference	4.32	4.07	4.18	4.11
Blockiness 5 × 5	1.77	2.09	1.93	2.11
Blockiness 8 × 8	1.37	1.52	1.49	1.54
Blurring ITU(F1)	2.57	2.67	2.56	2.64
Blurring Filter7	1.74	1.93	1.85	1.96
Jerkiness 6*Frames*	2.44	2.94	2.41	2.90
Jerkiness 11*Frames*	2.09	2.56	2.15	2.62
NoiseQ 3%	2.33	2.42	2.42	2.34
NoiseQ 15%	1.78	1.95	1.71	2.02
H.264 Bitrate 28kbps	1.34	1.45	1.08	1.49
H.264 Bitrate 56kbps	1.99	2.01	1.85	1.90
Packet Loss 0.5%	2.99	2.82	2.92	2.77
Packet Loss 1.5%	2.08	1.74	2.22	1.75
Lum. Impair. I (darker)	2.72	3.05	2.57	3.03
Lum. Impair. II (lighter)	2.83	3.08	3.02	3.07

subjective ratings from the VQDIM II study were used. Here, the test material was also *head and shoulder* video sequences as shown in Fig. 3.1. In this test, much more different test conditions are taken into account, as well as combinations of test conditions. Moreover, the test participants were also different from those used for model training. For more details about the VQDIM II experiment refer to Sect. 4.4.2.

In Fig. 5.2, the result of the modeling is shown. One can see that there is also a robust linear relationship between the subjective video quality ratings and video quality modeled from the dimension scores. In three cases, the model underestimates the quality score is this way, that it falls below the MOS scale range (*Bitrate28 kbps, Jerkiness 6 Frames + Blurring Filter1 and NoiseQ3 + Packet Loss 1.2%*). In these cases, the model sets the scores to 1. The correlation is very strong and the coefficient is $r_{model-VQDIM\ II} = 0.93$ and the RMSE is 0.34. It can be seen that the linear regression model provides reliable results with a high correlation and a low error. Furthermore, it should be underlined that the test conditions that were not in the initial model training data are modeled with high accuracy.

Fig. 5.1 Comparison of the subjective quality ratings from the two experiments (*VQDIM I, SD-PCA*) with the predicted MOS from the linear model (*MOS$_{pre-lin}$*) using the five perceptual dimension scores gathered in the VQDIM I experiment

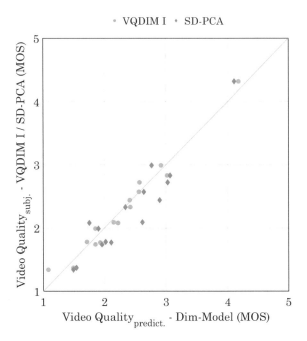

5.1.4   Quality Modeling—Other Video Content—VQDIM III

This section gives a brief look into the capability of the linear model to model the video quality for broader video content. The linear quality model was applied to the subjective ratings to predict the overall video quality. As described in Sect. 4.5, the applicability of the five video quality dimension was investigated. The experiment investigates a broad range of video content. This was in contrast to the *Head and Shoulder* video content in all the other experiments. Additional differences in the test material were that the videos have HD resolution and an aspect ratio of 16 : 9.

The model showed a high accuracy before, and therefore it is of interest to explore whether it achieves a similar accuracy in a broader video context. The findings could lay down the basis for the possible general applicability of the methods applied. The results of the overall video quality prediction (diamond marked graph) are shown together with the subjective quality ratings (dot marked graph) in Fig. 5.3.

It shows that, in general, the quality prediction works well. The correlation coefficient is $r = 0.99$. Nevertheless, the quality model underestimates the quality of all conditions. On average, the negative offset is 0.65 MOS. This result means that the quality model created in the videotelephony domain estimates the quality about half an MOS worse than they actually are. When correcting the predicted quality scores with the average offset of .65 MOS, the quality scores are almost identical with the subjective scores (see asterisk marked graph in Fig. 5.3). A reason for the constant underestimation could be that the model was trained with material that had

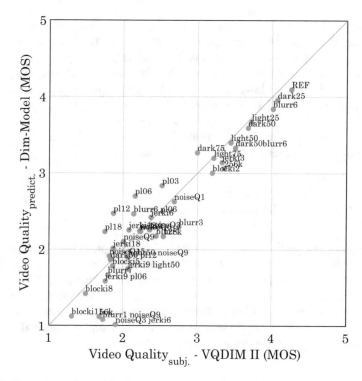

Fig. 5.2 Comparison of the subjective ratings from the experiment VQDIM II with the predicted MOS from the linear model ($MOS_{pre-lin}$)

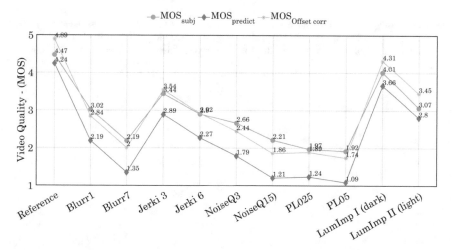

Fig. 5.3 Subjective MOS (dot marked graph) in comparison with predicted MOS (diamond marked graph) and offset corrected predicted MOS (asterisk marked graph) for overall video quality

a resolution of 640 × 480 pixel, and the test material in this study was presented in full HD. Therefore, the test participants could have rated the material better since it was larger when shown on the screen. This idea is backed by the finding in [1], where the author found that wider screen size can override other impairing effects (e.g., from codecs like H.264). On the other hand, the artifacts are probably more visible when presenting the video probes on a larger screen. It is unclear whether the offset is an effect of the model, the test group, or only appears within this data set. Nevertheless, it should be investigated in future studies, if the model should be updated with a parameter regarding the video resolution.

5.1.5 Quality Modeling—Conclusion

This chapter presented a linear model for video quality prediction from perceptual video quality dimension ratings. The evaluation of the model performance was done by comparing the results of the model with several subjective data. In general, the model based on linear regression does an excellent job of modeling the quality ratings from the dimension scores. Since the results are satisfying, the model can be further used when predicting the overall video quality from instrumental predicted dimension ratings, as presented later (see Chap. 6).

Reference

1. Jumisko-Pyykko, S.: I would like to see the subtitles and the face or at least hear the voice: effects of picture ratio and audio video bitrate ratio on perception of quality in mobile television. Multimedia Tools Applications, p. 2008. LLC, Springer Science and Business Media (2008)

Chapter 6
Instrumental-Dimension-Based Video Quality Modeling

6.1 Instrumental Dimension Prediction

6.1.1 Introduction

In this section, an approach to estimate the perceptual video quality dimensions from instrumentally obtained scores is presented. To create estimation models for each video quality dimension, the Video Quality Indicators (VQI), developed by the Video Quality Group at the AGH University of Science Krakow (comp. Sect. 2.7.4), was used.

From the vast set of indicators provided by the VQI, 10 were regarded as useful for the instrumental estimation and taken into account. These are

- *Blockniess, SpatialActivity, TemporalActivity, Block Loss, Blurring, Freezing, Exposure, Contrast, Interlace, Flickering*

The indicators *Exposure* and *Contrast* were converted before further usage. This step converted the two indicators in full-reference indicators. Instead of a linear scale, the amount of the difference of the impaired (Imp) to the reference (Ref) video was used:

$$\text{Exposure} = |\text{Exposure}_{\text{Imp}} - \text{Exposure}_{\text{Ref}}| \tag{6.1}$$

$$\text{Contrast} = |\text{Contrast}_{\text{Imp}} - \text{Contrast}_{\text{Ref}}| \tag{6.2}$$

All metrics are calculated frame by frame first and second aggregated by averaging the values over all frames. In addition, a difference value is calculated. Here, the lowest and the highest values for each metric in one frame is taken, and the difference is calculated. As before, the difference values are then averaged over all frames. This leads to a set of 20 indicators values for one test file, the average and difference value indexed as $ind X_{avg}$ and $ind X_{dif}$.

© The Author(s), under exclusive license to Springer Nature Switzerland AG 2021
F. Schiffner, *Dimension-Based Quality Analysis and Prediction
for Videotelephony*, T-Labs Series in Telecommunication Services,
https://doi.org/10.1007/978-3-030-56570-1_6

Because this set was not able to describe the dimension UCL to a sufficient extent, an additional metric was developed. This metric is designed as a full-reference metric and named *UCLmetric*. Here, for each frame, the difference in the power density spectrum between the impaired and the reference video was calculated. More information about the calculation of the *UCLmetric* is given in the next subsection.

6.1.1.1 UCLmetric

As said, this metric is meant to be used for the estimation of the Unclearness video quality dimension. Because it is a full-reference metric, it is necessary for the calculation that additionally to the impaired video, the reference signal is available.

The algorithm follows the following steps:

1. Find the belonging video pairs (impaired and reference). For each video-pair:
2. Grouping of frames. For each frame-pair:
3. Preprocessing;
4. Metric calculation per frame;
5. Calculation of the average of all frame results;
6. Write log file with results.

Matching the video pairs and grouping frames. The algorithm checks if the impaired and reference video is present. Afterward, both video signals are read, and the frames having the same index are grouped. The frames are handed over to the preprocessing.

Preprocessing. The preprocessing is applied to each frame separately, and the following processing steps are conducted:

1. The frame is converted to grayscale.
2. The frame is normalized.

   – Through that the metric becomes independent from luminance changes in the impaired video.

3. The frame is then filtered with a *medianBlur* filter.[1]

   – By that the *Salt-and-Pepper Noise* is eliminated.

4. A second filter is applied *GaussianBlur*.[2]

   – In this way, additional image noise and artifacts are reduced.

---

[1]The function smoothes an image using the median filter. Taken from—OpenCV (Open Source Computer Vision Library) https://docs.opencv.org/2.4/modules/imgproc/doc/filtering.html#medianblur last view: 02.12.2019.

[2]The function convolves the source image with the specified Gaussian kernel. Taken from—OpenCV (Open Source Computer Vision Library) https://docs.opencv.org/2.4/modules/imgproc/doc/filtering.html#medianblur last view: 02.12.2019.

5. (optional) a Laplace high-pass filter can be applied.

   – By that, the metric considers changes at the edges in a frame.

6. The frame is tailored to achieve a better run time for the FFT.

   – The frame is brought to the next best size for the FFT via zero padding. Afterward, only the first quadrant is used, since in the case of a "natural" scene a normal distribution of edges occur in the video. Thus, all quads are nearly equal.

Metric calculation. To calculated the metric, both frames are transformed with a Fast Fourier Transformation (FFT) into the frequency domain. Subsequently, both frames (Imp and Ref) are reduced to the first quadrant. The formula for the calculation is as follows:

$$
V_f = \frac{\sum_{i=1}^{I} \sum_{j=1}^{J} \max\left(E_{i,j}^{Ref} - E_{i,j}^{Imp}, 0 \right)}{\sum_{i=1}^{I} \sum_{j=1}^{J} E_{i,j}^{Ref}}
\tag{6.3}
$$

I–Width of the discrete spectrum
J–Height of the discrete spectrum
$E_{i,j}$–Energy of the discrete spectrum at i, j
f–Frame index

The results for each frame are stored in a list. After processing all frames of the video file, the average is determined. The final score is then stored in a log file for further usage.

Critical consideration. It should be noted that the metric is susceptible to Packet Loss. As said, the metric is developed as a full-reference metric, and here both, the original and the impaired, recorded video sequences are needed. The recorded PL-affected video sequences are shorter in terms of recorded frames. A time alignment was not possible because, from the recorded video file, it was not apparent which frames were lost. So, the time difference leads to a mismatch in the compared frames and distorts the result slightly. Typically the imagery does not change so rapidly from one frame to the next, so the error would not of enormous influence. Nevertheless, if the mismatch is precisely on a scene cut, then the error in estimation would be severe. Further, the higher the PL rate, the higher the mismatch of the compared frames. Knowing that, by future usage of the metric, the preprocessing should check for that the same frames are compared.

6.1.2 Determination of the Dimension Estimators

The first step was to calculate all video quality indicators and the *UCLmetric* for the used data sets. To create the models for each perceptual video quality dimension, polynomial regression with forward feature selection was used. The procedure looks for the best fitting indicators to depict the target dimensional score. The boundary conditions were that the dimensionality should be 1 or 2 and that the number of features used in the final modeling function should be between 2 to at max 6. To avoid overfitting in the model, several measures were taken. Two data sets were used (the ratings from the AV-DIM and VQDIM II experiments). The regression analysis was done for both data sets separately and tested with the other opposed. Further, the regression was conducted with a *train-test-split* setup, and the forward feature selection was conducted, using a fivefold cross-validation.

6.1.3 Estimation Models

The best performing models were chosen. Crucial for the choice was a good fit for the training and test data. Moreover, the fit should be good for both data sets, as well. At the same time, a focus was put on the lowest number of features and complexity of the resulting models. In the following, the models, as well as the performing results for both data sets, are given.

Estimation of "Fragmentation". The model that fits best for FRA estimation consists of three parameters in a linear combination. Since, PL introduces block artifacts into the image the VQI of *Blockiness* is used two times. Both the difference value and the average value is used to calculate the dimension score. Further, through the missing information of the lost packets, rapidly occurring blocks and slices introduce more *Spatial Activity*. To reflect that, the corresponding VQIs are used.

$$FRA = 3.66 + (-0.63 \cdot \text{Blockiness}_{\text{dif}})$$
$$+ (-0.32 \cdot \text{SpatialActivity}_{\text{dif}}) \qquad (6.4)$$
$$+ (-0.24 \cdot \text{Blockiness}_{\text{avg}})$$

Correlation on the training set $r = 0.70$
Correlation on the test set $r = 0.74$
Results
RMSE on the VQDIM II data set $= 0.56$
RMSE on the AV-DIM data set $= 0.47$

In both cases, on the training set and test set, the correlation shows a strong positive linear relationship. When looking at Fig. 6.1, one can see that the test conditions with a high score, meaning *no* or *little* Fragmentation are estimated well. In contrast, the test conditions that are rated below 3 on the scale by the test participants are over-

Fig. 6.1 Results of the Fragmentation (FRA) estimation. Comparison of subjective ratings and estimated scores for all 43 test conditions of the VQDIM II dataset

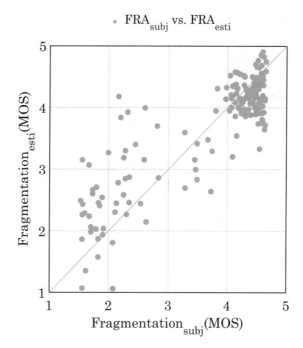

estimated. In some cases, the over-estimation is severe, meaning a miss-match up to 2 steps on the MOS scale.

Estimation of "Unclearness". Here, the model heavily relies on the *UCLmetric*. Additionally, the average of indicators *Blockiness* and *Blur* are used. Since block artifacts introducing areas of the same brightness and color value, the details of the imagery are lost. So, the resulting usage of this indicator is logical, as well as the *Blur* indicator.

$$UCL = 3.35 + (-0.93 \cdot \text{UCLmetric}_{\text{avg}})$$
$$+ (-0.27 \cdot \text{Blockiness}_{\text{avg}} \cdot \text{Blur}_{\text{avg}}) \tag{6.5}$$

Correlation on the training set $r = 0.79$
Correlation on the test set $r = 0.80$
Results
RMSE on the VQDIM II data set $= 0.40$
RMSE on the AV-DIM data set $= 0.24$

Regarding the performance of the model, the correlation is strong, and the error is relatively low. This is also reflected in Fig. 6.2, where the estimation compared to the subjective rating is shown.

Estimation of "Discontinuity". This dimension is estimated by using *Freezing* and *Flickering*. The *Freezing* indicator is coupled with the *Temporal Activity*. Therefore, only *Freezing* was used here, and the additional use of *Temporal Activity* was regarded

Fig. 6.2 Results of the
Unclearness (UCL)
estimation. Comparison of
subjective ratings and
estimated scores for all 43
test conditions of the
VQDIM II dataset

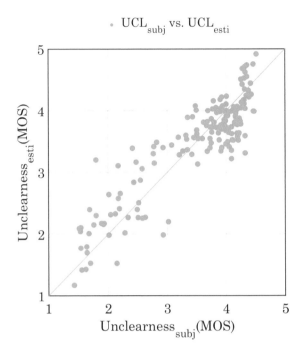

as redundant. The *Flickering* indicator was selected by the Forward Feature Selection
as the next best fitting one.

$$DIC = 3.72 + (-0.45 \cdot \text{Freezing}_{\text{avg}})$$
$$+ (-0.39 \cdot \text{Freezing}_{\text{dif}}) \qquad (6.6)$$
$$+ (-0.33 \cdot \text{Flickering}_{\text{dif}})$$

Correlation on the training set $r = 0.80$
Correlation on the test set $r = 0.72$
Results
RMSE on the VQDIM II data set $= 0.47$
RMSE on the AV-DIM data set $= 0.34$

The accuracy of the model shows similar results as the other models. In Fig. 6.3,
the comparison of the estimated dimension scores with the subjective Discontinuity
ratings is shown. One can observe that on the lower end of the scale, the values are
estimated well. On the upper end of the scale, one can observe that the estimation
is poor. The subjective score spread from 3 to almost 5 on the MOS scale, whereby
contrast, the estimated scores almost entirely lay by 4.2 on the MOS scale.

Estimation of "Noisiness". The best fitting model only uses the *Spatial Activity* and
Temporal Activity indicator. The intended Video Quality Indicators *Noise* does not
improve the estimation, quite the opposite. This indicator can only be used in our

Fig. 6.3 Results of the Discontinuity (DIC) estimation. Comparison of subjective ratings and estimated scores for all 43 test conditions of the VQDIM II dataset

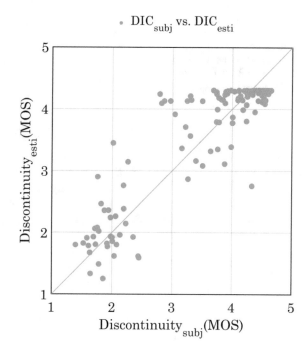

data set to tell if there is noise or not. The range of the results for detected noise was so small that to differentiate between the amount of noise in the sequence was not possible.

$$NOI = 3.80 + (-0.33 \cdot \text{SpatialActivity}_{avg})$$
$$+ (-0.57 \cdot \text{TemporalActivity}_{avg}) \tag{6.7}$$

Correlation on the training set $r = 0.69$
Correlation on the test set $r = 0.80$
Results
RMSE on the VQDIM II data set $= 0.57$
RMSE on the AV-DIM data set $= 0.40$

The correlation of the model is borderline strong on the training set and strong on the test set. This is due to the fact that the test conditions without noise are estimated quite well, in contrast to the subjective scores from test conditions affected by noise. Here, all subjective scores fall below about 2.5 MOS, but the estimation ranges up to roughly 3.5 MOS (Fig. 6.4).

Estimation of "Suboptimal Luminosity". This dimension can be estimated using only two measures. The indicator *Exposure* indicates how bright or how dark an image is. The indicator *Contrast* measures the relative variation of the luminance. The

Fig. 6.4 Results of the
Noisiness (NOI) estimation.
Comparison of subjective
ratings and estimated scores
for all 43 test conditions of
the VQDIM II dataset

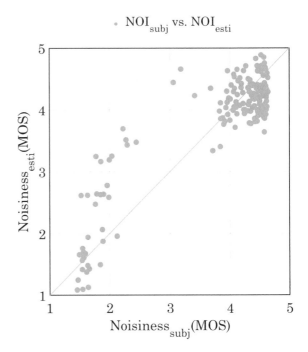

average of a video sequence is used from both indicators to estimate the dimension
rating.

$$LUM = 3.57 + (-0.5852 \cdot \text{Exposure}_{avg})$$
$$+ (-0.358 \cdot \text{Contrast}_{avg}) \quad (6.8)$$

Correlation on the training set $r = 0.96$
Correlation on the test set $r = 0.88$
Results
RMSE on the VQDIM II data set $= 0.43$
RMSE on the AV-DIM data set $= 0.21$

The performance of the model is the best of all five dimension-estimation models
with a high correlation and a small error. This can also be observed by looking at
Fig. 6.5.

Summarizing. All models showed a lower error on the AV-DIM data set as on the
VQDIM II data set. The dimension estimation with models constructed by using the
VQI does work okay but not as good as intended. Nevertheless, promising first steps
are presented.

There were several other metrics tested. The usage of metrics like SSIM, PSNR,
or the measures from the VMAF [1] quality estimator, did not improve the fit.

Fig. 6.5 Results of the Suboptimal Luminosity (LUM) estimation. Comparison of subjective ratings and estimated scores for all 43 test conditions of the VQDIM II dataset

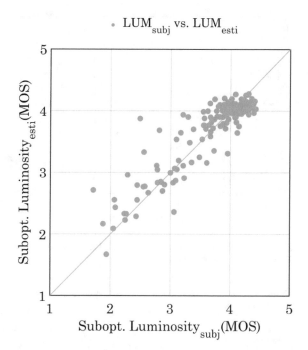

6.2 Dimension-Based Estimation of Overall Video Quality

In this section, the results of the overall video quality modeling are presented.

As summarized in the last section, the estimation of each video quality dimension can be regarded as good, but still lacks accuracy. Nevertheless, the results of the video quality dimension estimation are used to calculate the overall video quality. Therefore, the estimated dimension scores were fed to the linear video quality model (comp. Sect. 5.1) to calculate the video quality. The data set from the VQDIM II experiment was chosen because this data set contains the largest amount of test conditions. The final results are obtained by averaging over each test condition and are shown in Fig. 6.6 and Table 6.1. In Fig. 6.6 on the right side, one can see the comparison to the subjective obtained overall video quality ratings. At first glance, one can see that the model fit is good, and the ratings spread around the reference line. Nevertheless, one can also observe that the video quality scores obtained from the estimated video quality dimensions are slightly underestimated. This is especially true for the upper end of the quality scale. In order to be able to better classify the results, a comparison of the subjective overall video quality ratings with the video quality ratings calculated from the subjective dimension scores is given on the left side.

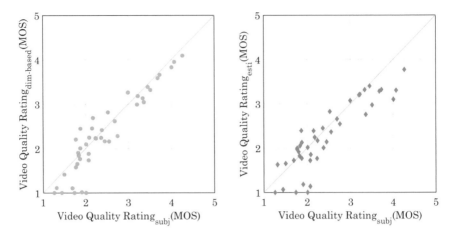

Fig. 6.6 Comparison of subjective ratings with dimension-based scores (left) and subjective ratings with estimated scores (right)

Table 6.1 Results from the video quality modeling (comparison of subjective vs. dimension-based vs. estimated video quality scores)

Test condition	VQ(subj.)	VQ(dimb.)	VQ(esti.)
Reference	4.26	4.09	3.77
Blockiness 2 × 2	3.20	3.00	3.20
Blockiness 5 × 5	1.87	1.77	2.40
Blockiness 8 × 8	1.50	1.42	1.65
Blockiness 11 × 11	1.32	1.11	1.63
Blurring ITU(F6)	4.01	3.83	3.10
Blurring ITU(F3)	2.75	2.28	2.55
Blurring ITU(F1)	2.45	2.16	2.14
Blurring Filter7	1.80	1.65	1.91
Jerkiness 3 Frames	3.33	3.14	3.32
Jerkiness 6 Frames	2.38	2.42	2.01
Jerkiness 9 Frames	2.23	2.23	1.77
Jerkiness 12 Frames	2.08	2.25	1.85
Jerkiness 18 Frames	1.87	2.01	1.77
NoiseQ 1%	2.69	2.62	2.66
NoiseQ 3%	2.36	2.24	2.45
NoiseQ 6%	2.24	2.23	2.17
NoiseQ 9%	2.01	2.10	2.02
NoiseQ 15%	1.82	1.90	1.82

(continued)

Table 6.1   (continued)

Test condition	VQ(subj.)	VQ(dimb.)	VQ(esti.)
H.264 Bitrate 256 kbps	3.36	3.05	2.76
H.264 Bitrate 128 kbps	2.55	2.16	2.37
H.264 Bitrate 56 kbps	1.69	1.11	1.73
H.264 Bitrate 28 kbps	1.26	1.00	1.00
Packet Loss 0.3%	2.52	2.82	2.82
Packet Loss 0.6%	2.17	2.69	2.24
Packet Loss 1.2%	1.88	2.45	2.13
Packet Loss 1.8%	1.77	2.20	2.00
LumImp I 25 darker	4.07	3.95	3.31
LumImp I 50 darker	3.68	3.59	3.29
LumImp I 75 darker	3.00	3.27	3.08
LumImp II 25 lighter	3.73	3.66	3.32
LumImp II 50 lighter	3.45	3.39	3.40
LumImp II 75 lighter	3.22	3.19	3.22
Blurring ITU F1 + NoiseQ 9%	1.74	1.08	0.76
Blurring ITU F6 + Packet Loss 0.6%	2.15	2.46	2.39
LumImp I 50 + Packet Loss 1.2%	1.84	1.84	2.05
LumImp I 50 + Blurring ITU F6	3.51	3.32	2.97
LumImp II 50 + NoiseQ 9%	2.07	1.88	1.14
Jerkiness 9 + LumImp II 50	2.07	1.74	1.72
Jerkiness 6 + Blurring ITU F1	2.01	1.00	1.00
Jerkiness 9 + Packet Loss 0.6%	1.77	1.58	1.97
NoiseQ 3% + Jerkiness 6	1.91	1.01	1.18
NoiseQ 3% + Packet Loss 1.2%	1.44	1.00	1.06

Despite the differences visible in the figure, the performance is quite similar for both:

subjective VQ versus dim-based VQ $\Rightarrow$ RMSE $= 0.47$; correlation $r = 0.93$
subjective VQ versus estimated VQ $\Rightarrow$ RMSE $= 0.43$; correlation $r = 0.86$

The fit of the modeled scores is very good with a high correlation and a low error. Nevertheless, knowing that underlying scores are predicted with less accuracy, it is assumed that the deviations in each dimension partially cancel each other out in the overall quality modeling process. This has a random component and suggests that the modeling of estimated dimension values of a completely unknown dataset does not yield such good results.

Reference

1. Netflix. VMAF-Video Multi-Method Assessment Fusion. https://github.com/Netflix/vmaf. Internet: Last view 10/2019

Chapter 7
Audio and Video Quality Modeling

7.1 Introduction

It is indisputable that for videotelephonynot only the video is important but also the transmitted speech signal. This chapter presents the investigation about the relationship between the perceptual quality dimensions of the video signal as well as the speech signal.

As described in Sect. 2.5.3, the approach of deconstructing the overall quality in its underlying perceptual dimensions was already done for speech telephony. The resulting perceptual speech quality space consists of four dimensions. Since, in the context of this work, the speech channel in video telephony is commonly named audio channel, the notion of audio dimensions is used, meaning the perceptual quality dimension describing *speech*. The prediction of audio-visual quality is commonly done by applying models (e.g., ITU-T- G.1070 [1], ITU-T- P.1201 [2]). The relationship between audio quality and video quality has been already studied intensively (comp. Sect. 2.3.4). This chapter aims at a better understanding of the interaction of the quality dimensions of the two perceptual spaces. The perceptual video quality dimensions and the perceptual audio quality dimension are shown in Table 7.1, together with a brief description.

7.2 Audio-Visual Quality Dimension Test

To investigate the interaction of the two perceptual quality spaces, an experiment was designed (audio-visual dimension test—AV-DIM). As before, the data set was used in this experiment, where short sequences of simulated videotelephony are shown. In the sequences, stories were told, such as a doctor's call or kitchen purchase [3] (comp. Sect. 3.1.1). Unlike the studies mentioned before, the audio channel was

Table 7.1 Description of the perceptual quality dimensions for video (VQD) and audio (AQD)

VQD	Name	Description
I	*Fragmentation (FRA)*	Fallen apart, torn and disjointed video
II	*Unclearness (UCL)*	Unclear and smeared image
III	*Discontinuity$_{video}$ (DIC$_v$)*	Interruptions in the flow of the video
IV	*Noisiness$_{video}$ (NOI$_v$)*	Random change in brightness and color
V	*Suboptimal Luminosity (LUM)*	Too high or low brightness
AQD	Name	Description
I	*Coloration (COL)*	Unnatural, metallic sound of the voice
II	*Discontinuity$_{audio}$ (DIC$_a$)*	Interruption in the flow of the speech sig.
III	*Noisiness$_{audio}$ (NOI$_a$)*	Unwanted hissy sound, added to the sig.
IV	*Loudness (LOU)*	Too high or low sound pressure

Table 7.2 Overview of test material

Material	Videotelephony/Head-and-Shoulder Scene
Number of speakers	4 persons (2 male/2 female)
Background settings	2 (office/living room)
File durations	approx. 10 s
No. of test conditions	35 + Reference
Resolution	640 × 480 Pixel
Frame rate	25 fps
Screen size (diagonal)	18.5 cm (7.3 inch)
Viewing distance	ca. 60 cm
Audio level	Individual most comfort level

active. The aim was to cover a broad range of potential degradations and mainly focuses on potential transmission degradations. To introduce degradations into the video, the Reference Impairment System for Video (RISV) [4] was employed. Also, the effects of coding and packet loss were included. As for the video channel, the audio channel was impaired with potential degradations that are common in speech transmission. To introduce impairments in the audio channel, the Software Tool Library (STL) provided by the ITU-T was employed [5]. Three classes of impairment were processed. First, only the audio channel was impaired and played together with the reference video. Second, only the video channel was impaired and played together with the reference audio. Third, both channels were impaired. It was taken into account that the impairments stimulate all perceptual dimensions. Furthermore, each combination of a video and an audio dimension are included in the test. For more details on the test material refer to Tables 7.2 and 7.3. In addition to the impairments in Table 7.3, the related video and audio dimensions are specified.

Table 7.3 Transformed quality ratings—Q (overall quality), VQ (video quality), AQ (audio quality)

#	Video Impair.	Audio Impair.	DIM_v	DIM_a	Q	VQ	AQ
00	Reference	Reference	–	–	4.18	4.30	3.96
01	Reference	Level −16 dBov	–	LOU	3.40	4.19	2.64
02	Reference	Level −36 dBov	–	LOU	4.17	4.27	4.03
03	Reference	G.722.2 PL0.02%	–	DIC_a	3.81	4.13	3.37
04	Reference	G.722.2 PL0.08%	–	DIC_a	3.06	4.08	2.03
05	Reference	G.711	–	COL	3.05	4.08	2.03
06	Reference	G.722.2 (6.6 kbs)	–	COL	3.80	4.12	3.53
07	Reference	MNRU-45	–	NOI_a	4.12	4.32	4.02
08	Reference	MNRU-55	–	NOI_a	4.12	4.34	3.82
09	Luminance (dark)	Reference	LUM	–	2.83	2.18	4.00
10	Luminance (light)	Reference	LUM	–	3.21	2.70	3.90
11	Blurring (Filter 1)	Reference	UCL	–	2.90	2.38	3.79
12	Jerkiness (6 Frames)	Reference	DIC_v	–	2.93	2.50	3.67
13	NoiseQ 3%	Reference	NOI_v	–	2.73	2.10	3.99
14	Packet Loss 1.5%	Reference	FRA	–	2.34	2.00	2.87
15	Luminance (dark)	Level −36 dBov	LUM	LOU	2.91	2.22	3.86
16	Luminance (dark)	G.722.2 PL0.08%	LUM	DIC_a	2.33	1.97	2.99
17	Luminance (light)	G.722.2 (6.6 kbs)	LUM	COL	2.84	2.43	3.26
18	Luminance (light)	MNRU-45	LUM	NOI_a	3.13	2.50	3.83
19	Blurring (Filter 1)	Level −36 dBov	UCL	LOU	2.84	2.38	3.80
20	Blurring (Filter 1)	G.722.2 PL0.08%	UCL	DIC_a	2.50	2.22	2.76
21	Blurring (Filter 1)	MNRU-45	UCL	NOI_a	2.60	2.12	3.53
22	Blurring (Filter 1)	G.711	UCL	COL	2.09	2.04	2.15
23	Jerkiness (6 Frames)	Level −16 dBov	DIC_v	LOU	2.55	2.38	2.73
24	Jerkiness (6 Frames)	G.722.2 PL0.08%	DIC_v	DIC_a	2.46	2.52	2.38
25	Jerkiness (6 Frames)	G.722.2 (6.6 kbs)	DIC_v	COL	2.59	2.54	2.57
26	Jerkiness (6 Frames)	MNRU-45	DIC_v	NOI_a	2.90	2.53	3.54
27	NoiseQ 3%	Level −16 dBov	NOI_v	LOU	2.31	2.03	2.61
28	NoiseQ 3%	G.722.2 PL0.08%	NOI_v	DIC_a	2.16	1.97	2.41
29	NoiseQ 3%	G.722.2 (6.6 kbs)	NOI_v	COL	2.39	1.94	3.05
30	NoiseQ 3%	MNRU-45	NOI_v	NOI_a	2.54	2.03	3.49
31	Packet Loss 1.5%	Level −16 dBov	FRA	LOU	1.98	1.83	2.20
32	Packet Loss 1.5%	G.722.2 (6.6 kbs)	FRA	COL	1.97	1.81	1.92
33	Packet Loss 1.5%	MNRU-45	FRA	NOI_a	2.20	1.85	2.60
34	Packet Loss 1.5%	G.711	FRA	COL	1.73	1.76	1.79

Fig. 7.1 7-point continuous scale, labels shown here are translated from German to English (corresponding values and German scale labels in brackets): extremely bad (1/extrem schlecht), bad (2/schlecht), poor (3/dürftig), fair (4/ordentlich), good (5/gut), excellent (6/ausgezeichnet), and ideal (7/ideal)

7.2.1 Test Participants and Procedure

A summary of the test participants can be found in Sect. 3.1.3. The test duration was roughly 55 min with an optional break of 5 min to avoid fatigue. In the beginning, a written introduction was given, and training was performed to allow the participants to familiarize themselves with the rating tasks and the degradations. The order of the test files was randomized for each participant to avoid the potential danger of order effects. The participants were able to replay the sample as often as needed. The test participants were allowed to adjust the volume in the beginning to the individual MCL. The rating task was divided into three parts. In the first part, the participants were asked to rate the overall quality, the video and audio quality of the video samples via a 7-point continuous scale (see Fig. 7.1) [6]. In the second and third parts, the participants should rate to what extent the quality-relevant dimensions for audio and video are present in the samples. The order of parts two and three was alternated between test participants. Moreover, the order of the rating scales in part two and three are randomized for each participant. The names of the dimensions were used as titles, and antonym pairs were used to describe the range of the scales.

In summary, the method of DSCAL was used for the video quality dimensions as well as for the audio quality dimensions, as described in Sect. 4.3.

7.2.2 AV-DIM—Data Analysis

The quality scores and the dimension scores from the experiment were analyzed and transformed into ratings of 5-point MOS scale as described in [7]. In general, it can be observed that the impairments intended to point to a corresponding perceptual dimension, indeed triggers them. The results for the quality ratings are shown in Table 7.3, and the results for the dimension ratings are shown in Table 7.4.

Table 7.4 Transformed dimension ratings for all perceptual video (DIM_v) and audio (DIM_a) dimensions

#	Video Impair.	Audio Impair.	FRA_v	UCL_v	DIC_v	NOI_v	LUM_v	COL_a	NOI_a	DIC_a	LOU_a
00	Reference	Reference	4.49	4.34	4.42	4.42	4.36	4.33	3.73	4.47	4.34
01	Reference	Level −16dBov	4.44	4.36	4.41	4.34	4.33	4.03	3.80	4.43	1.84
02	Reference	Level −36dBov	4.44	4.37	4.45	4.48	4.32	4.20	4.24	4.36	4.14
03	Reference	G.722.2 PL0.02%	4.46	4.22	4.35	4.43	4.27	3.78	4.05	3.64	4.13
04	Reference	G.722.2 PL0.08%	4.36	4.16	4.29	4.32	4.26	3.33	3.59	2.01	3.95
05	Reference	G.711	4.38	4.06	4.36	4.24	4.10	2.34	2.33	3.99	3.53
06	Reference	G.722.2 (6.6kbs)	4.41	4.17	4.38	4.41	4.22	3.51	3.94	4.32	4.16
07	Reference	MNRU-45	4.54	4.46	4.50	4.49	4.42	4.27	4.09	4.41	4.11
08	Reference	MNRU-55	4.52	4.40	4.50	4.45	4.31	4.13	4.02	4.44	3.78
09	Luminance (dark)	Reference	4.40	3.58	4.36	4.34	1.49	4.18	4.08	4.45	4.10
10	Luminance (light)	Reference	4.48	3.87	4.35	4.30	1.72	4.06	4.17	4.40	4.04
11	Blurring (Filter 1)	Reference	4.18	2.12	4.24	4.04	3.79	3.81	3.93	4.30	4.09
12	Jerkiness (6 Frames)	Reference	4.01	3.67	2.46	4.09	3.88	3.76	3.93	3.96	3.99
13	NoiseQ 3%	Reference	4.39	3.43	4.30	1.91	3.60	4.16	4.18	4.42	4.23
14	Packet Loss 1.5%	Reference	2.08	3.39	2.66	3.68	3.89	3.12	3.64	3.58	3.90
15	Luminance (dark)	Level −36dBov	4.46	3.47	4.36	4.28	1.45	4.17	4.28	4.38	4.19
16	Luminance (dark)	G.722.2 PL0.08%	4.35	3.07	4.14	4.14	1.40	3.38	3.31	3.33	3.84
17	Luminance (light)	G.722.2 (6.6kbs)	4.38	3.60	4.28	4.15	1.66	3.30	3.62	4.17	3.90
18	Luminance (light)	MNRU-45	4.41	3.79	4.42	4.26	1.68	4.05	4.00	4.42	4.16
19	Blurring (Filter 1)	Level −36dBov	4.36	2.08	4.27	4.09	3.62	3.95	3.99	4.37	4.02

(continued)

Table 7.4 (continued)

#	Video Impair.	Audio Impair.	FRA_v	UCL_v	DIC_v	NOI_v	LUM_v	COL_a	NOI_a	DIC_a	LOU_a
20	Blurring (Filter 1)	G.722.2 $PL0.08\%$	4.34	1.94	4.04	3.94	3.49	3.39	3.59	2.87	3.98
21	Blurring (Filter 1)	MNRU-45	4.22	1.95	4.19	4.00	3.65	3.88	3.78	4.27	3.75
22	Blurring (Filter 1)	G.711	4.08	1.92	4.05	3.77	3.70	2.44	2.58	3.92	3.71
23	Jerkiness (6 Frames)	Level -16dBov	4.11	3.65	2.30	4.09	4.01	3.77	3.51	4.15	1.81
24	Jerkiness (6 Frames)	G.722.2 $PL0.08\%$	3.86	3.49	2.36	4.12	3.89	3.09	3.70	2.56	3.87
25	Jerkiness (6 Frames)	G.722.2 (6.6kbs)	4.01	3.68	2.33	4.03	4.08	2.88	3.46	3.07	3.83
26	Jerkiness (6 Frames)	MNRU-45	4.08	3.70	2.42	4.09	4.01	3.83	3.73	3.90	4.01
27	NoiseQ 3%	Level -16dBov	4.15	3.40	4.10	1.73	3.65	3.79	3.44	4.16	2.01
28	NoiseQ 3%	G.722.2 $PL0.08\%$	4.14	3.17	3.89	1.72	3.46	3.26	3.48	2.75	3.96
29	NoiseQ 3%	G.722.2 (6.6kbs)	4.25	3.35	4.07	1.95	3.67	3.05	3.45	3.80	3.80
30	NoiseQ 3%	MNRU-45	4.19	3.48	4.22	1.80	3.62	3.90	3.45	4.30	4.00
31	Packet Loss 1.5%	Level -16dBov	1.78	3.36	3.02	3.62	3.72	3.06	3.39	3.70	1.71
32	Packet Loss 1.5%	G.722.2 (6.6kbs)	1.95	3.32	2.53	3.53	3.72	2.69	3.25	1.98	3.56
33	Packet Loss 1.5%	MNRU-45	1.80	3.38	2.87	3.61	3.83	2.97	3.60	3.22	3.92
34	Packet Loss 1.5%	G.711	1.81	3.08	2.55	3.59	3.85	2.40	2.75	2.23	3.55

7.2.3 Audio Quality and Prediction

In general, it can be observed that the impairments, as intended, trigger the corresponding dimension (e. g. the to loud audio condition *Level-16 dBov* had a definite impact on the Suboptimal Loudness dimension rating). Concerning the quality rating, the test participants rating behavior was as expected. The stronger the impairment the lower the quality score (e. g. audio Packet Loss condition 03 and 04 in Table 7.4). Furthermore, one can see that different degradations in the video channel did not impact the rating on the audio dimensions.

The only exceptions were the ratings when packet loss was introduced in the video. In that process, the audio channel was also/additionally impaired. On average, the standard deviation for the rating of the audio quality dimensions is low ($\sigma = 0.24$), smaller than a quarter MOS. The averaged standard deviation for the rating of the audio quality itself is also low ($\sigma = 0.29$). From the results is concluded, that the rating behavior on the audio quality dimension scales is consistent and therefore independent from the video channel and the video impairments.

The perceptual dimension-based audio quality prediction model is taken from [8]. There, the model was developed only in a speech setting and using linear regression. The formula is as follows:

$$MOS_{\text{audio}-\text{predict}} = -3.38 + (0.282 \cdot COL) + (0.467 \cdot NOI_a)$$
$$+0.7(\cdot DIC_a) + (0.345 \cdot LOU) \tag{7.1}$$

The audio quality modeling also works quite well in an audio-visual context, and the results are given in Table 7.6 and Fig. 7.2. The calculated PEARSON correlation is 0.97 and underlines the strong linear connection between the audio quality ratings from the subjects with the scores calculated from the dimension ratings. Also, the low RMSE of 0.26 shows the high accuracy of the model.

Result 1: From the results it is concluded that the audio quality can be predicted from the underlying audio quality dimensions and that the different video impairments do not influence the rating on the audio quality dimensions.

7.2.3.1 Comparison of Audio Dimension Ratings

In this section, a brief comparison between audio dimension ratings in a speech-only context with ratings in an audio-visual context is presented. The conditions used for that comparison are taken from [9] and are *G.711, G.722.2(23.05), G.722.2-PL0.2* and *G.722.2-PL0.8*. The test material used there were short sentences spoken in German with an average duration of $10s$. When regarding test conditions that are also part of a speech-only study, a high correlation between the ratings can be observed (see Table 7.5). In particular the audio dimension Coloration and Noisiness, here the

Table 7.5 Comparison of the subjective ratings for audio impairments for the dimensions COL, NOI, and DIC, in audio (speech) only and audio-visual context

MOS_{COL}	Audio-visual	Audio only	PEARSON
G.711	2.34	2.45	
G.722.2(23.05)	3.51	3.91	
G.722.2-PL0.2	3.78	4.16	
G.722.2-PL0.8	3.33	3.80	0.99
MOS_{NOI}	Audio-visual	Audio only	
G.711	2.33	3.10	
G.722.2(23.05)	3.94	4.66	
G.722.2-PL0.2	4.05	4.42	
G.722.2-PL0.8	3.59	4.60	0.94
MOS_{DIC}	Audio-visual	Audio only	
G.711	3.99	4.44	
G.722.2(23.05)	4.32	4.54	
G.722.2-PL0.2	3.64	1.92	
G.722.2-PL0.8	2.01	1.39	0.83

is correlation is > 0.9. The audio dimension Discontinuity shows a similar behavior. However, here, the condition $G.722.2 - PL0.02$ (marked red in Table 7.5) is out of the ordinary, which was rated significantly better in the audio-visual context. Here, it can be assumed that sometimes "advantageous" positions in the speech channel were impaired (for example, pauses in speech) so that the impairment cannot be heard. Because only 4 samples in the AV test were degraded with this impairment, one cannot make a definitive statement here. For the audio dimension LOU, there are no subjective dimension ratings for these test conditions in a speech-only setting. Nevertheless, the results indicate that speech quality dimensions are also stable in the AV context, and one can achieve very similar results.

7.2.4 Video Quality and Prediction

The same effects as in the audio channel can be observed as expected, like impairments in the video leads to a loss in video quality. Further, the "right" dimensions are triggered by the corresponding video degradations. It can be stated that the video quality dimensions are rated independently from the audio and the audio impairments. It can be observed that the ratings of the video quality did not change with the audio impairments.

The video quality rating stays nearly the same for one type of impairment, even if different audio impairments are present. No significant differences in the video ratings were found concerning the influence of the audio impairment. Furthermore, one can observe that an impairment in the video channel had a significant negative influence on the quality rating. Here, the video quality rating always dropped roughly below 2.5. The reason for that is that only test conditions are included in the experiment that clearly impair the video and trigger the video quality dimensions. This was done because more finely graduated test conditions would overexert the test participants. As a result, an "on-off" rating behavior was obtained, e. g. when the test participant spotted a video impairment, it leads subsequently to a low-quality rating.

The model for video quality prediction, as presented in Sect. 5.1, is used, and the formula is as follows:

$$MOS_{video-predict} = -5.94 + (0.21 \cdot FRA) + (0.73 \cdot UCL)$$
$$+ (0.53 \cdot DIC_v) + (0.47 \cdot NOI_v) + (0.31 \cdot LUM) \tag{7.2}$$

The results are compared to the subjectively obtained video quality scores and are shown in Table 7.6 and Figure 7.2. The calculated PEARSON correlation is 0.97 and underlines the strong linear connection between the video quality rating from the subjects with the score calculated from the video dimension scores. Also, the RMSE of 0.49 shows good accuracy of the model. This analysis shows that the video quality rating, as well as the video quality dimension ratings, stays the same when different audio impairments are present. Moreover, it could predict the video quality from the video quality dimension scores with good accuracy.

Result 2: The results lead to the conclusion that the video quality dimension, as well as the video quality prediction via the video quality dimensions, can be used in an audio-visual context.

7.2.5 Overall Quality Prediction

To model the overall quality, the approach that the audio and video quality is constituted separately and combined later on is used as described in Sect. 2.3.4. Therefore, the formula from [10] was used. The scaling factors are set to $\alpha = 1.5$ and $\beta = 0.121$ first. These values are the proposed maximal values given in [10].

$$MOS_{av} = \alpha + \beta \cdot MOS_a \cdot MOS_v \tag{7.3}$$

First, the overall quality from the ratings of the video quality and audio quality was calculated using Eq. (7.3). Second, the video quality and audio quality from the quality dimension ratings were calculated using Eqs. (7.2) and (7.1). Third, step the overall quality is calculated using the audio and video quality ratings calculated from the dimension ratings. The last step was to calculate the overall audio-visual quality with the help of Eq. (7.3). The PEARSON correlation between the subjective and the predicted MOS was calculated. In both cases, the correlation is very strong

Table 7.6 Subjective ratings and the results for the audio, video, and overall quality estimation, calculated by using the direct audio and video quality ratings and the dimension ratings

Video impairment	Audio impairment	MOS_a subj. Q	MOS_a predicted	MOS_v subj. Q	MOS_v predicted	MOS AV_{rate}	MOS AV_{rate}	MOS $DIM_a + DIM_v$
Reference	Reference	3.96	4.21	4.30	3.94	4.18	3.56	3.30
Reference	Level-16dBov	2.64	3.27	4.19	3.89	3.40	2.84	2.67
Reference	Level-36dBov	4.03	4.26	4.27	3.99	4.17	3.58	3.31
Reference	G.722.2 PL0.02%	3.37	3.55	4.13	3.78	3.81	3.19	2.97
Reference	G.722.2 PL0.08%	2.03	2.00	4.08	3.63	3.06	2.50	2.43
Reference	G.711	2.03	2.38	4.08	3.51	3.05	2.50	2.42
Reference	G.722.2 (6.6kbits)	3.53	3.91	4.12	3.73	3.80	3.26	3.00
Reference	MNRU-45	4.02	4.24	4.32	4.13	4.12	3.60	3.29
Reference	MNRU-55	3.82	4.07	4.34	4.03	4.12	3.51	3.25
Luminance (dark)	Reference	4.00	4.24	2.18	2.41	2.83	2.56	2.38
Luminance (light)	Reference	3.90	4.18	2.70	2.69	3.21	2.78	2.58
Blurring (Filter 1)	Reference	3.79	3.95	2.38	1.80	2.90	2.59	2.41
Jerkiness (6Frames)	Reference	3.67	3.67	2.50	2.00	2.93	2.61	2.42
NoiseQ3%	Reference	3.99	4.30	2.10	1.77	2.73	2.51	2.33
Packet Loss 1.5%	Reference	2.87	3.05	2.00	1.31	2.34	2.19	2.12
Luminance (dark)	Level-36dBov	3.86	4.30	2.22	2.29	2.91	2.54	2.39
Luminance (dark)	G.722.2 PL0.08%	2.99	2.78	1.97	1.78	2.33	2.21	2.12
Luminance (light)	G.722.2 (6.6kbits)	3.80	4.04	2.43	2.33	2.84	2.46	2.35
Luminance (light)	MNRU-45	3.83	4.16	2.50	2.61	3.13	2.66	2.51
Blurring (Filter 1)	Level-36dBov	2.76	2.64	2.38	1.79	2.84	2.59	2.39

(continued)

Table 7.6 (continued)

Video impairment	Audio impairment	MOS_a subj. Q	MOS_a predicted	MOS_v subj. Q	MOS_v predicted	MOS AV_{rate}	MOS AV_{rate}	MOS $DIM_a +$ DIM_v
Blurring (Filter 1)	G.722.2 PL0.08%	3.53	3.76	2.22	1.45	2.50	2.24	2.18
Blurring (Filter 1)	MNRU-45	2.15	2.54	2.12	1.59	2.60	2.40	2.25
Blurring (Filter 1)	G.711	3.26	3.50	2.04	1.38	2.09	2.03	2.01
Jerkiness (6Frames)	Level-16dBov	2.73	2.86	2.38	1.97	2.55	2.28	2.21
Jerkiness (6Frames)	G.722.2 PL0.08%	2.38	2.34	2.52	1.80	2.46	2.23	2.16
Jerkiness (6Frames)	G.722.2 (6.6kbits)	2.57	2.52	2.54	1.98	2.59	2.29	2.22
Jerkiness (6Frames)	MNRU-45	3.54	3.55	2.53	2.06	2.90	2.58	2.41
NoiseQ3%	Level-16dBov	2.61	2.90	2.03	1.53	2.31	2.14	2.10
NoiseQ3%	G.722.2 PL0.08%	2.41	2.45	1.97	1.18	2.16	2.08	2.04
NoiseQ3%	G.722.2 (6.6kbits)	3.05	3.06	1.94	1.61	2.39	2.22	2.14
NoiseQ3%	MNRU-45	3.49	3.72	2.03	1.68	2.54	2.36	2.22
Packet Loss 1.5%	Level-16dBov	2.20	2.25	1.83	1.34	1.98	1.99	1.98
Packet Loss 1.5%	G.722.2 (6.6kbits)	1.92	1.51	1.81	1.05	1.97	1.92	1.96
Packet Loss 1.5%	MNRU-45	2.60	2.74	1.85	1.31	2.20	2.08	2.05
Packet Loss 1.5%	G.711	1.79	1.37	1.76	0.92	1.73	1.88	1.89
Pearson R			0.97		0.97		0.98	0.98
RMSE			0.26		0.49			
RMSE $\beta = 0.121$							0.36	0.51
RMSE $\beta = 0.154$							0.11	0.20

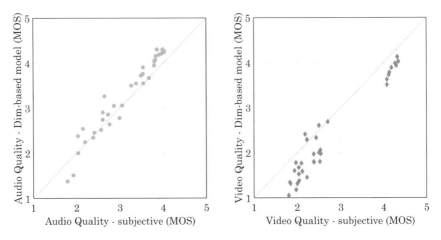

Fig. 7.2 Dimension-based quality prediction in comparison with the subjective quality rating for audio (left) and video (right)

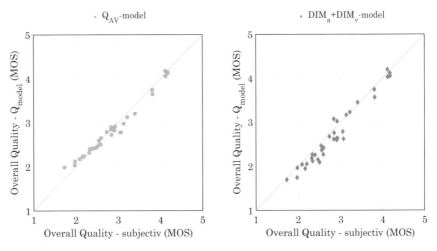

Fig. 7.3 Result of the audio-visual model (left side, dot-marked) and the combined dimension-based model (right side, diamond-marked)

with a coefficient of $r_{MOS_{subj.}\,versus\,MOS_{predict}} > 0.98$. While it is a meaningful measure to determine the consistency of the model, it does not give information about its accuracy. Therefore, the RMSE is more meaningful.

$$\text{RMSE}_{MOS_{subj.}\text{ versus }MOS_{AVpredict}} = 0.36$$
$$\text{RMSE}_{MOS_{subj.}\text{ versus }MOS_{DIMpredict}} = 0.51$$

This shows that the combination of the two linear regression models estimates the overall quality based on the nine (see Table 7.1) underlying perceptual dimensions with high reliability. However, it can be seen that this deviates from the quality modeling from only the video and audio quality rating in terms of accuracy. The overall quality scores obtained by combining the two models are slightly underestimated for $MOS < 2.5$ in comparison to the overall quality scores from the AV model. When comparing the results from both models with the overall quality rating directly obtained from the test participants, it can be observed that the overall quality is underestimated in almost all cases slightly. This holds accurate especially for the "better" conditions where the deviation is higher. The reason is that the scaling factor β does not fit for that case. Therefore, a new scaling factor β was calculated and set to 0.154.

This measure has enormously improved the accuracy of the prediction.

$$\text{RMSE}_{\text{MOS}_{\text{subj.}} \text{ versus MOS}_{\text{AVpredict}}} = 0.11$$
$$\text{RMSE}_{\text{MOS}_{\text{subj.}} \text{ versus MOS}_{\text{DIMpredict}}} = 0.20$$

The results from the overall audio-visual quality prediction using the new β scaling factor are shown in Fig. 7.3 and Table 7.6.

Result 3: Generally, the dimension-based quality modeling produces meaningful results and is nearly identical to the audio-visual model.

7.2.6 Conclusion

This experiment shows that the perceptual quality space for the video channel and the audio channel can be assessed directly via the corresponding perceptual quality dimensions in an audio-visual context. The two perceptual quality spaces are existing independently and do not interfere with each other. This also underlines the approach that the overall quality is a combination of the separately rated video quality and audio quality. The DSCAL method also works when video and audio are presented together. Further, it could be proven that the overall audio-visual quality can be predicted by using perceptual dimensions. The results lead to the suggestion that when dealing with audio-visual stimuli, both channels should be treated independently and combined later on to obtain an overall judgment.

References

1. ITU-T Rec. G.1070. Opinion Model for Video-Telephony Applications. International Telecommunication Union, CH-Geneva (2012)
2. ITU-T Rec. P.1201. Parametric non-intrusive Assessment of Audiovisual Media streaming Quality. International Telecommunication Union, CH-Geneva (2012)
3. Belmudez, B.: Audiovisual Quality Assessment and Prediction for Videotelephony. Springer, GER-Heidelberg (2015)
4. ITU-T Rec. P.930. Principles of a Reference Impairment System for Video. International Telecommunication Union, CH-Geneva 4/1996 (1996)
5. ITU-T Rec. G.191. Software Tools for Speech and Audio Coding Standardization. International Telecommunication Union, CH-Geneva (2010)
6. Raake, A.: Speech Quality of VoIP: Assessment and Prediction. Wiley, UK-Chichester (2006)
7. Köster, F. et al.: Comparison Between the Discrete ACR Scale and an Extended Continuous Scale for the Quality Assessment of Transmitted Speech. DAGA, GER-Nürnberg (2015)
8. Köster, F., Mittag, G., Möller, S.: Modeling the overall quality of experience on the basis of underlying quality dimensions. In: 2017 9th International Conference on Quality of Multimedia Experience (QoMEX 2017). IEEE -Signal Processing Society, GER-Erfurt (2017)
9. Wältermann, M.: Dimension-based Quality Modeling of Transmitted Speech. Springer, GER-Berlin (2013)
10. ITU-T Rec. P.911. Subjective Audiovisual Quality Assessment Methods for Multimedia Applications. International Telecommunication Union, CH-Geneva 12/1998 (1998)

Chapter 8
Instrumental Estimation of Dimension-Based Audio-Visual Quality

8.1 Instrumental Audio Quality Estimation with DIAL

This section presents the instrumental estimation of audio quality. Further, a comparison of the estimated scores with subjective obtained scores is given for each perceptual dimension.

In Sect. 2.7, the dimension-based model DIAL was described and is used in this section to estimate the quality-relevant dimensions for the speech signal in the audio channel. From the *AV-DIM* data set, the audio channel was extracted from the impaired test files and fed into the model. The respective audio reference signal was prepared as well and also fed to the model. In Fig. 8.1, the results for the four perceptual dimensions are shown.

The perceptual audio dimensions COL and DIC are estimated quite well. The correlation is strong with a coefficient of $r = 0.9$ and the RMSE is 0.46 for COL and, respectively, $r = 0.88$ and the RMSE is 0.34 for DIC. Therefore, one can say that these two dimensions are predicted with sufficient accuracy. The perceptual audio dimension NOI is estimated poorly. Almost all conditions are estimated massively better then the test subjects rated them. Here, the correlation is 0.51, and the RMSE is 0.88. This result was unexpected, but DIAL seems to be insensitive regarding noise. As results, all test conditions are estimated with a NOI-MOS of roughly about 4.4. When regarding the ratings for Suboptimal Loudness; here, all test conditions are predicted similar to the subjective rating besides one test condition. The test condition *Level -16dBov* is rated roughly opposite to the subjective ratings and are estimated as almost *excellent* when the test subjects rate them as *bad-poor*. This result was surprising and led to the suggestion that DIAL misinterprets "too loud" as a good thing. Nevertheless, this "misinterpretation" leads to a weak correlation and a huge RMSE. Taken that test condition aside, the model predicts the other test condition sufficiently. All results are also shown in Table 8.1.

© The Author(s), under exclusive license to Springer Nature Switzerland AG 2021
F. Schiffner, *Dimension-Based Quality Analysis and Prediction for Videotelephony*, T-Labs Series in Telecommunication Services, https://doi.org/10.1007/978-3-030-56570-1_8

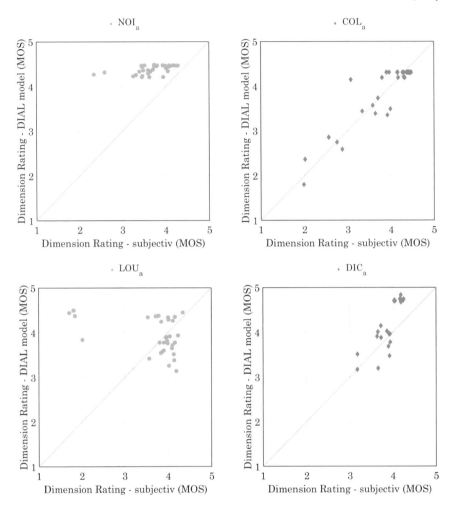

Fig. 8.1 Comparison of the subjective audio quality dimension ratings with the estimated scores from DIAL (COL—Coloration, NOI—Noisiness, LOU—Loudness, DIC—Discontinuity)

8.2 Synthesis—Instrumental-Dimension-Based Transmitted Speech and Video Quality

In this section, the results from the instrumental-dimension-based video quality and audio quality estimation will be combined. The aim is to estimate the overall quality of the audio-visual stimuli based on instrumental measures and underlying quality dimension estimation.

Therefore, the estimated audio dimension scores from the last section are fed into the linear model, as presented in Sect. 7.2. The same procedure was chosen for video quality dimensions. The video quality dimension scores are obtained by

Table 8.1 Subjective audio quality dimension ratings and the estimated audio quality dimension ratings (via DIAL)

Video impairment	Audio impairment	COL_a		NOI_a		LOU_a		DIC_a	
		subj.	esti.	subj.	esti.	subj.	esti.	subj.	esti
Reference	Reference	4.33	4.18	3.73	4.48	4.34	4.45	4.47	4.31
Reference	Level-16 dBov	4.03	4.18	3.80	4.47	1.84	4.37	4.43	4.30
Reference	Level-36 dBov	4.20	4.18	4.24	4.47	4.14	3.38	4.36	4.31
Reference	G.722.2 PL0.02%	3.78	3.93	4.05	4.34	4.13	3.51	3.64	3.39
Reference	G.722.2 PL0.08%	3.33	3.92	3.59	4.37	3.95	3.89	2.01	2.36
Reference	G.711	2.34	3.17	2.33	4.27	3.53	4.34	3.99	3.49
Reference	G.722.2 (6.6 kbits)	3.51	3.65	3.94	4.21	4.16	3.78	4.32	4.19
Reference	MNRU-45	4.27	4.05	4.09	4.45	4.11	4.26	4.41	4.31
Reference	MNRU-55	4.13	4.03	4.02	4.42	3.78	4.38	4.44	4.31
Luminance (dark)	Reference	4.18	4.25	4.08	4.47	4.10	3.72	4.45	4.31
Luminance (light)	Reference	4.06	4.18	4.17	4.47	4.04	3.91	4.40	4.31
Blurring (Filter 1)	Reference	3.81	4.17	3.93	4.47	4.09	3.65	4.30	4.31
Jerkiness (6Frames)	Reference	3.76	4.17	3.93	4.47	3.99	3.76	3.96	4.31
NoiseQ3%	Reference	4.16	4.16	4.18	4.47	4.23	3.93	4.42	4.31
Packet Loss 1.5%	Reference	3.12	3.63	3.64	4.33	3.90	3.60	3.58	3.57
Luminance (dark)	Level-36 dBov	4.17	4.16	4.28	4.47	4.19	3.14	4.38	4.31
Luminance (dark)	G.722.2 PL0.08%	3.38	3.86	3.31	4.25	3.84	4.24	3.33	3.44
Luminance (light)	G.722.2 (6.6 kbits)	3.30	3.65	3.62	4.21	3.90	3.78	4.17	4.19
Luminance (light)	MNRU-45	4.05	4.03	4.00	4.41	4.16	4.35	4.42	4.31
Blurring (Filter 1)	Level-36 dBov	3.95	4.18	3.99	4.46	4.02	3.26	4.37	4.31
Blurring (Filter 1)	G.722.2 PL0.08%	3.39	3.90	3.59	4.29	3.98	3.81	2.87	2.59
Blurring (Filter 1)	MNRU-45	3.88	4.03	3.78	4.45	3.75	4.37	4.27	4.31
Blurring (Filter 1)	G.711	2.44	3.17	2.58	4.32	3.71	4.37	3.92	3.35
Jerkiness (6Frames)	Level-16 dBov	3.77	4.23	3.51	4.47	1.81	4.50	4.15	4.31
Jerkiness (6Frames)	G.722.2 PL0.08%	3.09	3.94	3.70	4.37	3.87	3.58	2.56	2.85
Jerkiness (6Frames)	G.722.2 (6.6 kbits)	2.88	3.72	3.46	4.23	3.83	3.54	3.07	4.15
Jerkiness (6Frames)	MNRU-45	3.83	4.03	3.73	4.46	4.01	4.29	3.90	4.31
NoiseQ3%	Level-16 dBov	3.79	4.18	3.44	4.47	2.01	3.84	4.16	4.31
NoiseQ3%	G.722.2 PL0.08%	3.26	3.91	3.48	4.36	3.96	3.90	2.75	2.74
NoiseQ3%	G.722.2 (6.6 kbits)	3.05	3.65	3.45	4.21	3.80	3.78	3.80	4.19
NoiseQ3%	MNRU-45	3.90	4.03	3.45	4.44	4.00	4.34	4.30	4.20
Packet Loss 1.5%	Level-16 dBov	3.06	3.72	3.39	4.33	1.71	4.44	3.70	3.74
Packet Loss 1.5%	G.722.2 (6.6 kbits)	2.69	3.65	3.25	4.23	3.56	3.42	1.98	1.80
Packet Loss 1.5%	MNRU-45	2.97	3.62	3.60	4.31	3.92	4.26	3.22	4.20
Packet Loss 1.5%	G.711	2.40	3.08	2.75	4.33	3.55	4.24	2.23	2.98
Pearson R		0.9		0.53		0.88		−0.36	
RMSE		0.49		0.8		0.34		0.91	

Fig. 8.2 Comparison of subjective overall quality ratings with the combined estimated dimension-based quality scores

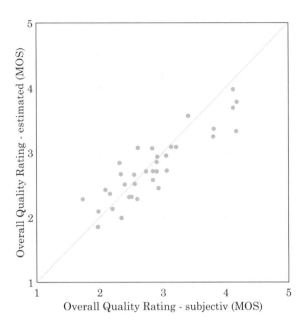

Table 8.2 Subjective overall quality ratings (subj.) and instrumental estimated scores (esti.) the audio- (AQ), video- (VQ), and overall quality (Q)

Video impairment	Audio impairment	$MOSsubj.Q$	$MOSesti.Q$	$MOSesti.VQ$	$MOSesti.AQ$
Reference	Reference	4.18	3.78	3.29	4.50
Reference	Level-16 dBov	3.40	3.57	3.39	3.96
Reference	Level-36 dBov	4.17	3.33	3.30	3.60
Reference	G.722.2 PL0.02%	3.81	3.37	3.47	3.50
Reference	G.722.2 PL0.08%	3.06	2.72	3.37	2.35
Reference	G.711	3.05	2.95	3.32	2.84
Reference	G.722.2 (6.6 kbits)	3.80	3.25	3.39	3.34
Reference	MNRU-45	4.12	3.69	3.39	4.20
Reference	MNRU-55	4.12	3.97	3.57	4.50
Luminance (dark)	Reference	2.83	3.07	2.51	4.06
Luminance (light)	Reference	3.21	3.09	2.57	4.01
Blurring (Filter 1)	Reference	2.90	2.85	2.19	4.01
Jerkiness (6Frames)	Reference	2.93	2.45	1.60	3.87
NoiseQ3%	Reference	2.73	2.71	1.95	4.02
Packet Loss 1.5%	Reference	2.34	1.99	1.01	3.17
Luminance (dark)	Level-36 dBov	2.91	2.94	2.56	3.64
Luminance (dark)	G.722.2 PL0.08%	2.33	2.67	2.45	3.10

(continued)

Table 8.2 (continued)

Video impairment	Audio impairment	$MOSsubj.Q$	$MOSesti.Q$	$MOSesti.VQ$	$MOSesti.AQ$
Luminance (light)	G.722.2 (6.6 kbits)	2.84	2.71	2.36	3.34
Luminance (light)	MNRU-45	3.13	3.09	2.42	4.27
Blurring (Filter 1)	Level-36 dBov	2.84	2.58	1.94	3.60
Blurring (Filter 1)	G.722.2 PL0.08%	2.50	2.32	2.08	2.55
Blurring (Filter 1)	MNRU-45	2.60	3.07	2.30	4.44
Blurring (Filter 1)	G.711	2.09	2.42	2.09	2.87
Jerkiness (6Frames)	Level-16 dBov	2.55	2.51	1.53	4.30
Jerkiness (6Frames)	G.722.2 PL0.08%	2.46	2.31	1.79	2.95
Jerkiness (6Frames)	G.722.2 (6.6 kbits)	2.59	2.28	1.60	3.18
Jerkiness (6Frames)	MNRU-45	2.90	2.71	1.79	4.39
NoiseQ3%	Level-16 dBov	2.31	2.84	1.99	4.37
NoiseQ3%	G.722.2 PL0.08%	2.16	2.36	1.95	2.87
NoiseQ3%	G.722.2 (6.6 kbits)	2.39	2.51	1.95	3.34
NoiseQ3%	MNRU-45	2.54	2.66	1.85	4.06
Packet Loss 1.5%	Level-16 dBov	1.98	2.09	1.00	3.84
Packet Loss 1.5%	G.722.2 (6.6 kbits)	1.97	1.85	1.01	2.28
Packet Loss 1.5%	MNRU-45	2.20	2.13	1.00	4.10
Packet Loss 1.5%	G.711	1.73	2.28	1.85	2.73
Pearson R			0.89		
RMSE			0.31		

using the estimation models presented in Chap. 6. The new calculated quality scores for audio and video quality are further combined using Eq. 7.3 given in Sect. 7.2.5. Here, the same newly calculated scaling factor $\beta = 0.154$ was used to increase the goodness of the fit. The result is shown in Fig. 8.2 and Table 8.2. The performance is good with a very strong correlation of $r = 0.89$. The error lies in the expected range (RMSE $= 0.31$). By choosing the new scaling factor β, it can be seen that the modeling is not perfect, but in general sufficient. The instrumental-dimension-based approach underestimates, especially at the upper end to the scale, the overall quality slightly.

Chapter 9
Conclusion and Future Work

This chapter summarizes the findings of this work and answers the research questions formulated in Sect. 1.2. Further, this work will close with an outlook of future work.

9.1 Summary

The research in this work focused on the perceptual quality of video transmission. The constant improvement of technology, the development of the Internet, smartphones, and the underlying technology, made videotelephony, and the transmission of video to almost every place at any time possible.

The theoretical background and related work are presented in Chap. 2. First, an introduction to perception is given and followed by the description of modern signal transmission. Here, also the main concepts of error recovery are explained. It is followed by a description of the concept of *Quality* and how judgment is formed. Third, theoretical issues about subjective testings are discussed. Further, different experimental approaches toward a multidimensional investigation of overall quality are explained: a Semantic Differential (SD) measures quality via predefined sets of Antonym Pairs (AP) and a Paired Comparison (PC) test evaluates the perceived similarity of a pair of stimuli. The chapter closes with a brief look into instrumental quality estimation for audio and video transmission.

In Chap. 3, all the details of the conducted experiments are laid out. Here, besides the experimental setup, the introduced approaches from the last chapter are employed to deconstruct the overall quality into a set of underlying quality-relevant perceptual dimensions. Finally, the chapter ends with the setup of the perceptual video quality space, consisting of five dimensions.

In Chap. 4, a new test method named Direct Scaling (DSCAL) is introduced. With the help of this method, the quality dimensions are rated directly by naïve test participants. The results of the three validation experiments were presented. Two experiments are conducted in the videotelephony domain, and one in a broader video setting.

The following next two chapters were about overall quality modeling. Chapter 5 presented a linear model that predicts the overall video quality from subjectively gathered quality dimension ratings. Chapter 6, however, presented an approach to estimate the video quality dimensions rating only by instrumental metric. Here, VQI are used to build estimation models for each quality dimension. The results from both chapters are compared to subjective overall video quality ratings, which represent the ground truth.

The exploration of the interaction of perceptual video quality dimensions with the perceptual audio quality dimension was presented in Chap. 7. An experiment using the DSCAL method for both modalities was presented, and the video, audio, and overall quality was modeled.

The second last Chap. 8 gave an insight into the instrumental estimation of audio quality dimensions using DIAL. All results from the chapters before were taken into account, and entirely instrumental obtained rating scores were compared to subjective ratings, and the accuracy of the estimation was given.

9.2 Conclusion

This section concludes the findings in this work and answers the research questions.

The first question is directly the core question, and is as follows:

What are the underlying perceptual quality dimensions for video?

The perceptual video quality dimensions are determined using mainly test material reflecting videotelephony. Nevertheless, the work from Tucker was also taken into account. There, the author used a broad range of video content, mainly from IPTV. This work underlines the idea that there may be a perceptual quality space for video in general.

From the experiment results, five underlying perceptual dimensions were determined (comp. Sect. 3.4):

1. Fragmentation (FRA) indicates to what extent the video image is incomplete and broken into parts.
2. Unclearness (UCL) relate to smeared imagery. Here edges and colors are not sharp and appear washed out.
3. Discontinuity (DIC) refers to interruptions in the flow of the videos.
4. Noisiness (NOI) indicates the amount of noise that can be found in the frames.
5. Suboptimal Luminosity (LUM) gives an insight into how much the luminosity deviates from the ideal. It can be either too bright or too dark.

Following these findings lead to the development of a sophisticated and new test method for video quality assessment (DSCAL). The method is not only easy to use for the test participants but also a reduction in experimental effort. The experiments conducted to validate the method were successful in gathering quality dimension ratings.

Using the results from the experiments, a linear regression model was trained to predict the overall video quality from the dimension ratings. This leads to the next question and the following answer:

> *Is it possible to model the overall video quality from the underlying video quality dimensions?*

The overall video quality can be estimated with the help of the five perceptual dimensions, as described in Sect. 5.1. The comparison of subjective overall quality ratings with the quality rating calculated from the dimension ratings showed that it is possible to predict the overall quality from only the dimension ratings.

The third research question was

> *Is it possible to estimate the ratings of each video quality dimension separately and model the overall video quality from the estimated dimension ratings?*

A first approach was used to estimate the video quality dimensions by the so-called Video Quality Indicators. Further, an own quality metric was presented. Five different models were calculated for each video quality dimension separately. Afterward, the video quality is calculated by using the beforehand determined dimension-based linear quality model. The results are promising, but not all perceptual dimensions are estimated with a sufficient accuracy (comp. Chap. 6). The ratings for the dimensions UCL and LUM are estimated best, followed by NOI. The models had problems to estimate the dimensions FRA and DIC. Nevertheless, the results when regarding the calculated overall video quality from the estimated video quality dimension ratings are good and showed the right direction.

Since typically video is transmitted together with the audio channel and it was muted until this point, the next research questions were as follows:

> *Is it possible to assess the audio and video quality dimensions at the same time, and how do they interact? Further, is it possible to combine an instrumental-dimension-based audio quality estimator with a dimension-based video quality estimator to predict the overall quality?*

The conducted study showed that the test participants could differentiate between the two perceptual modalities and their quality dimensions. The results showed that the overall, audio, and video quality could be calculated from linear-dimension-based models, as presented in Chap. 7. The results are suggesting that the audio and video channels are independently processed in higher cognitive areas. Further, the quality for each perceptual modality can be determined separately and fused later on to form the overall quality. When regarding the combined instrumental quality dimension estimation, the estimation can be regarded as good but still lacks accuracy.

9.3 Future Work

The conclusions from the last section allow multiple directions to future research questions that aim to further coloring the *map* mentioned in the opening.

The whole concept of perceptual Video Quality Dimensions could be investigated further in different settings. Since this work was done in a passive test scenario, it could be investigated to what extent the VQD are applicable in an active videotelephony setting. The question here could also be whether the conversation dimensions from speech telephony [1] can be used too.

Another direction could be to further generalize the VQD to video in general, including streaming, gaming, mobile video, motion pictures, or even virtual reality applications.

It should be noted that all the ratings are obtained from short video sequences (<15 s) and are, therefore, short-term judgments. The establishment of a long-term quality judgment is out of the scope but would be of interest. Furthermore, long-term changes in perceptual quality and the relationship to the underlying perceptual dimension would be a worthful future work item. The service provider could also be interested in how a bad long-term quality judgment can be recovered with the goal to regain lost users.

When regarding the modeling of the perceptual VQD, one can think of creating data sets from crowdsourcing or record real-life data and using the newest machine-learning techniques to predict each dimension.

Altogether, it should be pursued to further contribute to the ITU-T Study Group 12. Collaborating, but still independent, research facilities are needed to use, evaluate, and further develop these methods and models. This would take the whole *quality community* forward, especially in times with intense pressure from the market.

Furthermore, taking into account that technological advancement will continue, the perceptual video quality space should also be under constant observation. If a new technology emerges, it could bring new types of degradation that may not be covered yet. So it could be possible that new perceptual dimensions arise and existing dimensions fade. One should be aware of the fact that *quality* is not static, but an ever-moving construct.

"We are all Humans until
Race disconnected us,
Religion separated us,
Politics divided us,
And Wealth classified us!"
—Pravinee Hurbungs

Reference

1. Köster, F.: Multidimensional Analysis of Conversational Telephone Speech. Springer, GER-Berlin (2018)

Appendix A
Expert Survey and Pretest

In this section, the questionnaires for the expert survey and the pre-test are given. The questionnaires are prepared in German. The participants are allowed to answer in German and English if wanted.

© The Author(s), under exclusive license to Springer Nature Switzerland AG 2021   127
F. Schiffner, *Dimension-Based Quality Analysis and Prediction*
for Videotelephony, T-Labs Series in Telecommunication Services,
https://doi.org/10.1007/978-3-030-56570-1

Bestimmung des relevanten Vokabulars zur Beschreibung
von Videobeeinträchtigungen:

Vielen Dank, dass Sie sich die Zeit nehmen, um mich bei meiner Untersuchung zu unterstützen. Ziel der Arbeit ist es den subjektiven Wahrnehmungsraum für Videostörungen zu untersuchen.

Zu diesem Zweck müssen Attribute gefunden werden, die die wahrgenommenen Störungen entsprechend beschreiben oder generell den technischen Zustand des Videomaterials beschreiben.
Der Inhalt eines etwaigen Videos spielt hierfür keine Rolle, ausschließlich die von einem menschlichen Betrachter wahrgenommenen Beeinträchtigungen und die Anmutung sind hierfür von Interesse. Bitte bearbeiten Sie zuerst Fragen 1, bevor Sie das Blatt wenden und sich Frage 2 zu wenden.

1.) Bitte notieren Sie (in deutscher oder englischer Sprache):

Adjektive (*„natürlich"*)

Nomen (*„Natürlichkeit"*)

Antonym-Paare („natürlich - unnatürlich")

welche aus Ihrer Erfahrung heraus Videos und Videoübertragungen am besten beschreiben können. Gehen Sie dabei insbesondere auf nicht optimale Videoübertragung und Videobeeinträchtigungen ein. In dieser Untersuchung geht es nur um „das Bild". Bitte lassen Sie etwaigen Videoinhalt jeder Art und auch eventuelle Audioinhalte außer Acht.

_____

_____

_____

_____

_____

_____

_____

_____

_____

2.) Bitte wählen Sie aus der nachfolgenden Liste 10 Antonym-Paare aus, die Ihnen für die Bewertung von Videobeeinträchtigungen am wichtigsten/aussagekräftigsten erscheinen.

Setzen Sie dafür bitte ein Kreuz in die letzte Spalte.

Adjektiv	Beschreibung	Antonym	
bewegungsunscharf	Unschärfe bei Bewegungen im Bild	bewegungsscharf	
blockig	vereinzelte Blöcke im Bild	nicht blockig	
farbverzerrt	falsche, unnatürliche (unpassend im Zusammenhang) Farben	farbrichtig	
flimmernd	zittern, flimmern innerhalb des Bildes	nicht flimmernd	
infofehlend	Informationen im Bild/Bildteil fehlen	infovollständig	
kariert	kleine Blöcke im gesamten Bild	unkariert	
kontraststark	sehr starker Farbkontreast (unnatürlich stark)	kontrastschwach	
konturvorgehoben	unnatürlich starke Konturen innerhalb des Bildes	konturlos	
künstlich	unnatürlich erscheinendes Bild	natürlich	
matschig	Bild bereichsweise verschmiert (unscharf)	nicht matschig	
nachhallig	Geisterbilder, Nachhall von Bildteilen	nicht nachhallig	
ruckartig	kurzes und oft stoppendes Bild (stotternd)	flüssig	
schattig	Schatten von Objekten zu sehen	nicht schattig	
stagnierend	aufgehängt, länger anhaltendes Bild	nicht stagnierend	
streifig	vereinzelt Streifen im Bild	ungestreift	
überbelichtet	Bild blendet	unterbelichtet	
überlagert	Bildverschiebung, Bildüberlagerung (ganzes Bild)	nicht überlagert	
unscharf	Unschärfe im gesamten Bild	scharf	
verpixelt	kleine Blöcke bereichsweise erkennbar	einheitlich	
verrauscht	körnig erscheinendes Bild	rauschfrei	
wackelig	hin und her wackeln innerhalb des Bildes	fest	
wellig	wellig erscheinendes Bild	eben	
zerfasert	mehrere untereinander versetzte Streifen, Linienfehler	nicht zerfasert	
zerfließend	zerfließende Bildpunkte, Detailunschärfe	detailscharf	
zerstückelt	allgemeiner Eindruck von Stückelung im Bild	zusammenhängend	

SD-DIM-AT-PT

Bitte kreuzen Sie **alle** Wortpaare an, die die Videoprobe beschreiben können

Adjektiv		Antonym
bewegungsunscharf		bewegungsscharf
blockig		nicht blockig
farbverzerrt		farbrichtig
flimmernd		nicht flimmernd
konstraststark		kontrastschwach
künstlich		natürlich
matschig		nicht matschig
ruckartig		flüssig
stagnierend		nicht stagnierend
streifig		ungestreift
überbelichtet		unterbelichtet
überlagert		nicht überlagert
unscharf		scharf
verpixelt		einheitlich
verrauscht		rauschfrei
wackelig		fest
zerfasert		nicht zerfasert
zerstückelt		zusammenhängend

Bitte geben Sie, falls möglich, weitere Adjektive oder Attribute an, die Ihrer Meinung nach die Probe zusätzlich beschreiben könnten

| |
| |
| |
| |

List of attributes obtained in the expert survey:

jerky - smooth, blocky, grainy, blurry, distorted, dull (verblasst), verpixelt, flüssig, abgehackt, unscharf, klar, clarity, wasch out, freeze, Artefakte, Codec Artefakte, blocks, frame loss, slices, stottern, ruckeln, verpixelt - nicht verpixelt, springend - stabil, unscharf - scharf, verfärbt - echt, slicing, abgeschnitten, verschoben,

Blöcke, Auflösung, Ruckeln, blockig, verschmiert, blurry, Blockartefakte, Blockfehler, Skalensprünge (bei SVC), ruckelig, fehlerbehaftet, wackelig - ruhig, flickern - glatt, sprunghaft - flüssig, scharf - unscharf, verrauscht, schmierend, pixelig, unscharf, verrauscht, geblurred, motion blur, Verzerrung, zeitlich instabil, schlechter Kontrast, blockig, körnig, verwaschen, ruckelig, blass, unscharf, Farbrauschen, Sand, Gamma, Kontrast-Fehler, Bildfehler, scharf - unscharf, blass - überfärbt, farbintensiv - matt, frabstichig, fleckig, ruckelig, wolkige Bewegungen, zerschreddert, rauschig/verrauscht, eingeforen, sprunghafter Farbverlauf, detailverdeckend, blockhaft, stockend, kontrastarm, farbig - matt, detailiert, matschig,

Similar attributes are grouped and antonyms created. This resulted in a huge set of Antonym Pairs. The set was reduced and after a pre-test the most often used AP are used in the Semantic Differential (SD) experiment. The results of the pre-test is shown in Table A.1.

Table A.1 Antonym pairs from the pre-test and the amount that specific pair was chosen to be present in the prepared data set

Adjective	Antonym	Amount
bewegungsunscharf	bewegungsscharf	134
blockig	nicht blockig	97
farbverzerrt	farbrichtig	93
flimmernd	nicht flimmernd	78
konstrast stark	kontrast schwach	42
künstlich	natülich	101
matschig	nicht matschig	62
ruckartig	flüssig	115
stagnierend	nicht stagnierend	19
streifig	ungestreift	54
überbelichtet	unterbelichtet	20
überlagert	nicht überlagert	79
unscharf	scharf	116
verpixelt	einheitlich	131
verrauscht	rauschfrei	74
wackelig	fest	103
zerfasert	nicht zerfasert	74
zerstückelt	zusammenhängend	97

Appendix B
Test Instruction

In the following, an example of the test instruction for the Direct Scaling (DSCAL) method is given. The test instructions were altered slightly in the different experiments to meet the specific requirements.

© The Author(s), under exclusive license to Springer Nature Switzerland AG 2021
F. Schiffner, *Dimension-Based Quality Analysis and Prediction*
for Videotelephony, T-Labs Series in Telecommunication Services,
https://doi.org/10.1007/978-3-030-56570-1

**Test instructions for the Video Quality – Video Dimension Test
– translated from German**

<u>Assessment of the Video Quality and Video Properties</u>

Thank you for taking part in this experiment! Please take the time to read the instructions completely. Please, switch off your mobile phone now. Should you have any questions, please do not hesitate to ask the experiment supervisor.

You are now participating in an experiment to evaluate the quality and the properties of video samples. These are video samples that may contain different degradations.

The experiment is divided into two tasks. The following is the guide for the first task of the experiment. For the second task, there are separate instructions afterward.

<u>First Task - Assessment of video quality:</u>

You will then be presented with different video samples. These are excerpts of video-telephone calls. The samples are presented one after the other. After you have viewed a sample, please submit your assessment. In increments, you specify how your personal perceived overall impression is relative to the quality of the video sample. The thematic content of the video sample should not play any role in the evaluation. The descriptive adjectives on the rating scale are intended to assist you in your judgment. The sound is switched off in this experiment. The figure below shows the input window.

Input window: video quality:

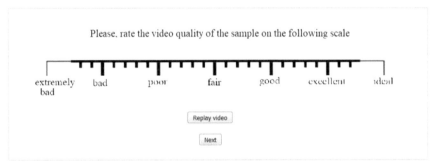

Before the start of the test, a training phase takes place, which gives you an overview of the degradations occurring in the experiment. In addition, the training phase allows you to familiarize yourself with the assessment task and the computer program. The training phase is displayed on the screen.

Please rate the sample after it has been played. To do this, click on the corresponding point on the scale with the mouse. After you set your rating, you can refine your rating on the scale if necessary.

If you wish, you can see the video again by clicking the "Replay video" button. Once you are done with your evaluation, click the "Next" button to go to the next rating task.

Second task – video properties :

In the following, you should evaluate the properties of the video samples. The samples shall be evaluated using **five descriptive scales.** The order of the scales may be different in the test than in the following explanation. The use of each scale will be explained in more detail below.

1.) Noisiness
The scale is called "Noisiness" and is labeled "noisy" and "noiseless". With "Noisiness" is meant how much noise is present in the video.

The scale is as follows:

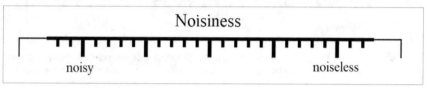

The term "noisy" can be described with terms like "noise" and "flickering". The term "noiseless" can be circumscribed by terms such as "not noisy" and "not flickering".

If you feel the video sample is very noisy put the cross to the following position:

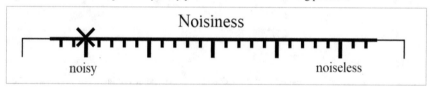

If you cannot detect any noise, place the cross in the "noiseless" position:

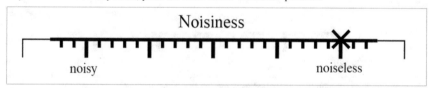

You can use the entire range of the scale to describe the degree of degradation. If you think the degree of degradation is only moderate, you could move the slider into this area:

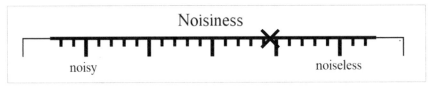

Perhaps the sample is clearly noisy, but not quite as extreme; then your evaluation might look as follows:

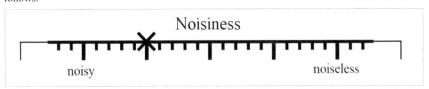

In principle, you can also use the spaces in between the markers, if necessary. In particular, you can use the "overflow areas" beyond the terms, if the terms for the assessment are not sufficient for you, e.g.:

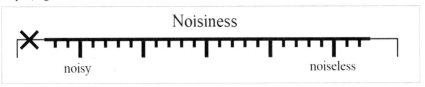

(2.) Unclearness

With the this scale, the "Unclearness" of the video sample is to be determined. Here is meant, like how blurred, unclear, or washed out a video picture is.

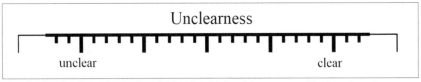

The term "unclear" can also be described as "muddy" and "contrast weak". "Clear" can be described as "not muddy" and "contrasting". The scale is used in the same way as for "Noisiness" (see above).

(3.) Discontinuity

The next scale is used to describe the "Discontinuity" of the sample.

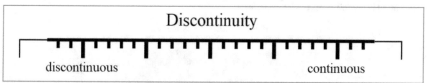

By the term "discontinuous" is meant that the video sample is played "jerky" and "wobbly". "Continuous" can be described as "constant" and "stable". The scale is used in the same way as for "Noisiness" (see above).

(4.) Fragmentation

The next scale is used to assess "Fragmentation". By "Fragmentation" it is meant how far the video breaks into individual parts, which means in individual fragments.

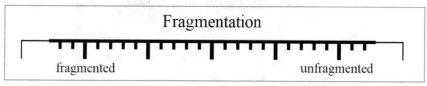

By the term "fragmented" is meant that the video sample is played "blocky" and "dismembered". "Unfragmented" can be circumscribed with the terms "non-blocking" and "contiguous". The scale is used in the same way as for "Noisiness" (see above).

(5.) Suboptimal Luminosity

The last scale is used to determine suboptimal luminosity. In this case, the video image can have an optimal luminosity or a non-optimal, i.e., suboptimal luminosity. It is irrelevant whether the luminosity is judged too dark or too light. In either case, it can be referred to as suboptimal.

The scale is used in the same way as for "Noisiness" (see above).

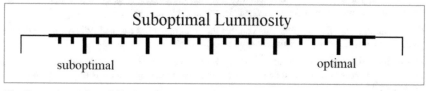

The five scales are hopefully already clear. In order to know how to rate the video image, you will see some typical video samples that can be assigned to one of the scales. Please, familiarize yourself with the scales, synonyms, and related video examples. Only if you are aware of the properties of the sample with the respective scale, read on. If you have any questions, please contact the test supervisor.

The sound is switched off in this experiment.

Example for the video player:

Example of the rating scale:

If you wish, you can see the video again by clicking the "Replay video" button. Once you are done with your evaluation, click the "Next" button to get to the next rating scale

Please continue until the program tells you that the experiment has finished. When you are done, please come out of the experiment room and report it to the experimenter.

Please make the assessment intuitive. In this investigation, which is purely subjective, there are neither correct nor false answers. Only your personal impression is important for the investigation.

Should you have any questions, please do not hesitate to ask the experiment supervisor.

Thank you for your participation!

Appendix C
File Processing

C.1 File Processing—MATLAB Code

Example from the MATLAB code written for RISV video processing below:

Listing C.1 MATLAB code RISV host script

```
1   %%%%%%%%%%%%%%%%%%%%%%%%%%%%%%%%%%%%%%%%%%%%%%%%%%%%%%%%%%%%%%%%%%%%%%
2   % HOST_RISV_Processing
3   %
4   % (c) Falk Schiffner M.Sc. Technische Universit\"at Berlin,
        Germany
5   %   06/2015 All Rights Reserved
6   %
7   % This Script allows to creat video impairments to video input
        files
8   % according to the ITU-T P.930 Rec.
9   %
10  %%%%%%%%%%%%%%%%%%%%%%%%%%%%%%%%%%%%%%%%%%%%%%%%%%%%%%%%%%%%%%%%%%%%%%
11  %%B E G I N   C O D E
12  %% HEAD
13  clc; clear all; close all
14
15  %% Add Source and Target Directories, Add "Function-Folder"
16      tmp = dir('ProcessedFilesNEW');
17      if isempty(tmp)
18          mkdir('ProcessedFilesNEW');
19      end
20
21      tmp = dir('VOIPrefNEW');
22      if isempty(tmp)
23          mkdir('VOIPrefNEW');
24      end
```

```
25
26      addpath('VOIPrefNEW/aavi');
27      addpath('Sourcefiles');
28      addpath('ProcessedFilesNEW');
29      addpath('FunctionsRISV');
30      clear tmp;
31
32   %% Reading Source Files Names
33      folder = 'VOIPrefNEW/aavi/*.avi';
34      videodata = dir(folder);
35
36   %% Input Processing Parameter
37      disp('--- Select desired Impairment Type --- ');
38      disp('---------------------------------------');
39      disp('artifical Blockiness    => ''block'' ');
40      disp('artifical Blurring      => ''blurr'' ');
41      disp('artifical Edge Busyness => ''edgeB'' ');
42      disp('artifical Noise (quant) => ''noiseQ'' ');
43      disp('artifical Jerkiness     => ''jerki'' ');
44      disp('---------------------------------------');
45      impairment = input('Please Input Impairment Type From The
            List Above: ');
46
47   %%
48   tStartOverall = tic;
49   switch impairment
50      %%
51      case 'block'
52          disp(' ');
53          disp('--- Artifical Blockiness Starts ---');
54          disp('Select Size Of Blocks (in Pixel):' )
55          blocki_value = input('1 = No Blockniess => 30 = Insane
                Big Blocks: ');
56          disp('Select Percent Of Frames To Be Blocki:' )
57          blocki_frame_percent = input('0 - 100%: ');
58          disp('Select Percent Of Blocks Within A Frame To Be
                Impaired:' )
59          block_percent = input('0 - 100%: ');
60
61          wait = waitbar(0,'Please Wait - File Processing ...');
62
63          for i = 1:length(videodata)
64              disp(['Processing File: ' num2str(i)]);
65              video_input = videodata(i).name;
66              tmpname = strrep(video_input, '.avi','');
67              output_name = ['ProcessedFilesNEW\' tmpname '_blocki_
                    ' , num2str(blocki_value),'_framesimp_',num2str(
                    blocki_frame_percent),...
68                  '_blocksimp_',num2str(block_percent) '.avi'];
69              [num_frames, num_rows, num_columns] = readVideoInput
                    (video_input);
70
```

```
71        [status] = artificalBlockiness(video_input,
             num_frames, ...
72                            num_rows, num_columns, blocki_value,
                               blocki_frame_percent,
                               block_percent, output_name);
73        disp(['Processing File: "', output_name ,'" ' status
             ]);
74        waitbar(i / length(videodata));
75    end
76    close(wait)
77
78  case 'blurr'
79      disp(' ');
80      disp('--- Artifical Blurring Starts ---');
81      disp('Select Methode for Blurring: ');
82      disp('Blurr via Moving Average: ''MovAve'' ');
83      disp('Blurr via ITU-T Rec.930 Filter: ''ITU930'' ');
84      blurrmeth = input('Methode: ');
85      switch blurrmeth
86          case 'MovAve'
87              disp('Select Smoothing Value (Nummer Of Pixel To
                   Average):' )
88              smooth_val = input('1 = No Blurring => xx =
                   Number of Pixel: ');
89          case 'ITU930'
90              disp('Select Degree of Impairment :' )
91              smooth_val = input('1 highest ==> 6 lowest: ');
92          otherwise
93              error('ERROR => Impairment Type Unknown ==> False
                   Input');
94      end
95
96      wait = waitbar(0,'Please Wait - File Processing ...');
97
98      for i = 1:length(videodata)
99          disp(['Processing File: ' num2str(i)]);
100         video_input = videodata(i).name;
101         tmpname = strrep(video_input, '.avi','');
102         output_name = ['ProcessedFilesNEW\' tmpname '_blurr_'
                , num2str(smooth_val), num2str(blurrmeth),'.avi'
                ];
103         [num_frames, num_rows, num_columns] = readVideoInput
                (video_input);
104         switch blurrmeth
105             case 'MovAve'
106                 [status] = Blurring_RowColumnsMovAverage(
                        video_input, output_name, num_frames,
                        num_rows, num_columns, smooth_val);
107             case 'ITU930'
108                 [status] = artificalBlurring(video_input,
                        output_name, smooth_val, num_frames,
                        num_rows);
109         end
```

```
110         disp(['Processing File: "', output_name ,'" ' status
                ]);
111         waitbar(i / length(videodata));
112     end
113     close(wait)
114
115 case 'noiseQ'
116     disp(' ');
117     disp('--- Artifical Quantization Noise ---');
118     disp('Select Percent Of Frames To Be Noisy:' )
119     procent_noise = input('0 - 100% (Note: 10% already very
                high Noise): ');
120
121     wait = waitbar(0,'Please Wait - File Processing ...');
122
123     for i = 1:length(videodata)
124         disp(['Processing File: ' num2str(i)]);
125         video_input = videodata(i).name;
126         tmpname = strrep(video_input, '.avi','');
127         output_name = ['ProcessedFilesNEW\' tmpname '_noiseQ_
                ' , num2str(procent_noise),'.avi'];
128         [num_frames, num_rows, num_columns] = readVideoInput(
                video_input);
129
130         [status] = artificalQuantNoise(video_input,
                output_name, procent_noise, num_frames, num_rows,
                num_columns);
131         disp(['Processing File: "', output_name ,'" ' status
                ]);
132         waitbar(i / length(videodata));
133     end
134     close(wait)
135
136 case 'edgeB'
137     disp(' ');
138     disp('--- Artifical Edge Busyness ---');
139     disp('Select Level of Impairment To Be Created:' )
140     val_res = input ('-1 ... -30: ');
141
142     wait = waitbar(0,'Please Wait - File Processing ...');
143
144     for i = 1:length(videodata)
145         disp(['Processing File: ' num2str(i)]);
146         video_input = videodata(i).name;
147         tmpname = strrep(video_input, '.avi','');
148         output_name = ['ProcessedFilesNEW\' tmpname '_edgeB_'
                , num2str(val_res),'.avi'];
149         [num_frames, num_rows, num_columns] = readVideoInput
                (video_input);
150
151         [status] = artificalEdgeBusyness(video_input,
                output_name, val_res, num_frames, num_rows,
                num_columns);
```

```
152            disp(['Processing File: "', output_name ,'" ' status
                   ]);
153            waitbar(i / length(videodata));
154        end
155        close(wait)
156
157     case 'jerki'
158        disp(' ');
159        disp('--- Artifical Jerkiness Starts ---');
160        disp('Select Nummer Of Frames To Skip:' )
161        holdframe = input('1 = No Skip => xx = Number of Frames:
                   ');
162
163        wait = waitbar(0,'Please Wait - File Processing ...');
164
165        for i = 1:length(videodata)
166            disp(['Processing File: ' num2str(i)]);
167            video_input = videodata(i).name;
168            tmpname = strrep(video_input, '.avi','');
169            output_name = ['ProcessedFilesNEW\' tmpname '_jerki_'
                   , num2str(holdframe),'.avi'];
170            [num_frames, num_rows, num_columns] = readVideoInput
                   (video_input);
171
172            [status] = artificalJerkiness(video_input, num_frames
                   , holdframe, output_name);
173            disp(['Processing File: "', output_name ,'" ' status
                   ]);
174            waitbar(i / length(videodata));
175        end
176        close(wait)
177
178     otherwise
179        error('ERROR => Impairment Type Unknown ==> False Input'
                   );
180 end
181
182 disp('File Processing Finished');
183 tEndOverall = toc(tStartOverall);
184 disp(['Processing Time: ' num2str(tEndOverall) 's']);
185
186 %%% E N D  O F  F I L E
```

Listing C.2 MATLAB code for artificial Blurring

```
1 function [status] = artificalBlurring(video_input, output_name,
      impair_level, num_frames, num_rows)
2 %%%%%%%%%%%%%%%%%%%%%%%%%%%%%%%%%%%%%%%%%%%%%%%%%%%%%%%%%%%%%%%%%%%
3 % articlaBlurring.m, v.2.1
4 % copyright - Falk Schiffner M.Sc., TU Berlin,
5 % Germany 07/2015 All Rights Reserved
6 % Creating Blurring Effect with regards to the ITU-T Rec. 930
7 %%%%%%%%%%%%%%%%%%%%%%%%%%%%%%%%%%%%%%%%%%%%%%%%%%%%%%%%%%%%%%%%%%%
```

```
 8   tStartblurrFrame = tic;
 9
10   video_in = VideoReader(video_input);
11   video_out = VideoWriter(output_name);
12   video_out.FrameRate = 25;
13   open(video_out);
14
15   %% Impulsrespones for diffenrent Impairment Level from ITU-T Rec
       . 930
16   imp_res_level_1 = [-1, 1, 3, 6, 10, 13, 15, 16, 15, 13, 10, 6,
       3, 1, -1];
17   imp_res_level_2 = [-2, -1, 1, 5, 9, 14, 17, 19, 17, 14, 9, 5, 1,
       -1, -2];
18   imp_res_level_3 = [-3, -3, -1, 3, 8, 15, 20, 22, 20, 15, 8, 3,
       -1, -3, -3];
19   imp_res_level_4 = [0, -3, -5, -3, 5, 15, 24, 28, 24, 15, 5, -3,
       -5, -3, 0];
20   imp_res_level_5 = [2, 1, -4, -6, -1, 13, 28, 34, 28, 13, -1, -6,
       -4, 1, 2];
21   imp_res_level_6 = [-2, 2, 4, -3, -9, 3, 31, 47, 31, 3, -9, -3,
       4, 2, -2];
22   % Added filter for "more" Blurring
23   imp_res_level_7 = [-1, 1, 3, 6, 6, 10, 10, 13, 15, 15, 16, 18,
       18, 20, ...
24   18, 18, 16, 15, 15, 15, 13, 10, 10, 6, 6, 3, 1, -1];
25       %%
26       if impair_level == 1
27           imp_res = imp_res_level_1;
28       elseif impair_level == 2
29           imp_res = imp_res_level_2;
30       elseif impair_level == 3
31           imp_res = imp_res_level_3;
32       elseif impair_level == 4
33           imp_res = imp_res_level_4;
34       elseif impair_level == 5
35           imp_res = imp_res_level_5;
36       elseif impair_level == 6
37           imp_res = imp_res_level_6;
38       elseif impair_level == 7
39           imp_res = imp_res_level_7;
40
41       else
42           error('ERROR => Level Of Impairment Incorrect')
43       end
44
45
46       zpad=zeros(1, length(imp_res));
47       front = ceil(length(imp_res)/2);
48       back = floor(length(imp_res)/2);
49
50       frame = double(read(video_in, 1));
51       frameycbcr = double(zeros(size(frame)));
52       frame_conv = double(zeros(size(frame)));
```

```
53    framergb = double(zeros(size(frame)));
54
55    [z, ~] = size(frame);
56    padding = zeros(z, length(zpad));
57    frame_conv_padded = [padding frame_conv(:,:,1) padding];
58
59
60    for f = 1 : num_frames
61        clear frame frame_block;
62        frame = double(read(video_in, f));
63
64        %% Convert Frame to YCBCR whole Frame at once
65        [frameycbcr] = convertRGBtoYCBCR_frame(frame);
66
67        zpad_Y_front = fliplr(frameycbcr(:, 1 : length(zpad), 1)
                );
68        zpad_Y_back = fliplr(frameycbcr(: , end - length(zpad) +
                1 : end, 1));
69        frame_ycbcr_padded = [zpad_Y_front frameycbcr(:, :, 1)
                zpad_Y_back];
70        %% CONVOLUTION WITH IMPULSRESPONSE Row by Row
71
72        for count_row = 1 : 1 : num_rows
73            frame_row_y_padded = frame_ycbcr_padded(count_row, :,
                    1);
74            row_tmp_y = (1 / (sum(imp_res)) * conv(double(imp_res
                    ), double(frame_row_y_padded)));
75            frame_conv_padded(count_row, :, 1) = row_tmp_y(front
                    : end - back);
76        end
77
78        frame_conv_cut(:, :, 1) = frame_conv_padded(:, 1 +
                length(zpad) : end - length(zpad), 1 );
79        frame_conv(:, :, 1) = frame_conv_cut(:, :, 1);
80        frame_conv(:, :, 2 : 3) = frameycbcr(:, :, 2 : 3);
81        %% Convert Frame Back to RGB
82        [framergb] = convertYCBCRtoRGB_frame(frame_conv);
83
84    writeVideo(video_out, uint8(framergb));
85    end
86    close(video_out);
87    status = 'finished';
88    tEndblurrFrame = toc(tStartblurrFrame);
89    disp(['Processing Time: ' num2str(tEndblurrFrame) 's']);
90 end
91 %%% E O F
```

Listing C.3 MATLAB code for AMR-WB audio processing

```
1 %%%%%%%%%%%%%%%%%%%%%%%%%%%%%%%%%%%%%%%%%%%%%%%%%%%%%%%%%%%%%%%%%%%
2 % (c) Falk Schiffner M.Sc. Technische Universitaet Berlin,
    Germany
3 %   11/2017 All Rights Reserved
```

```matlab
4   %
5   %%%%%%%%%%%%%%%%%%%%%%%%%%%%%%%%%%%%%%%%%%%%%%%%%%%%%%%%%%%%%%%%%%%%%%%
6   %%B E G I N    C O D E
7   %% HEAD
8   clc; clear all; close all
9
10  %% Add Source and Target Directories, Add "Function-Folder"
11      tmp = dir('ProcessedFiles2');
12      if isempty(tmp)
13          mkdir('ProcessedFiles2');
14      end
15
16      tmp = dir('SourceFiles');
17      if isempty(tmp)
18          mkdir('SourceFiles');
19      end
20
21      addpath('SourceFiles_PL_AVI' , 'SourceFiles');
22      addpath('ProcessedFiles' , 'ProcessedFiles2');
23      addpath('AMRWB_PLC');
24      addpath('TestFileProcessing');
25      clear tmp;
26
27  %% Reading Source Files Names
28      folder = 'SourceFiles/*.avi';
29      videodata = dir(folder);
30
31  %% Input Processing Parameter
32      disp('--- START AMRWB Processing --- ');
33      disp('------------------------------');
34
35  %%
36  tStartOverall = tic;
37  PLval = input('Please Input Percentage of PL you want to
        introduce (2% -> 0.02): ');
38  wait = waitbar(0,'Please Wait - File Processing ...');
39
40          for i = 1:length(videodata)
41              disp(['Processing File: ' num2str(i)]);
42              video_input = videodata(i).name;
43              tmpname = strrep(video_input, '.avi','');
44              output_name = ['ProcessedFiles2\' tmpname '_AudioPL_'
                    , num2str(PLval), '.avi'];
45              [num_frames, num_rows, num_columns] = readVideoInput
                    (video_input);
46
47              [status, deg, ref, errFlag, actualErrRate, t, fs] =
                    amrwbPLC(video_input, PLval);
48              tmpaudiofilename = [tmpname '_AudioPL_' , num2str
                    (PLval), '.wav'];
49              audiowrite(tmpaudiofilename, deg, fs);
50
51              video_source = ['SourceFiles\',video_input];
```

```
52      audio_source = tmpaudiofilename;
53
54      Outputfilename = [tmpname '_G7222_PL',num2str(PLval),
            '.avi'];
55
56      ffmpeg_string_AVmerge = ['ffmpeg -i ', video_source ,
            ' -i ' ,audio_source ,' -c:v copy -c:a copy -map
            0:v:0 -map 1:a:0 ' , ' AV', Outputfilename];
57      dos(ffmpeg_string_AVmerge);
58
59      mergedFilename = ['AV', Outputfilename];
60      target = ['ProcessedFiles2\', mergedFilename];
61      movefile(mergedFilename, target ,'f');
62
63      delete *.wav
64      disp('AUDIO VIDEO MERGE d o n e');
65
66
67      disp(['Processing File: "', output_name ,'" ' status
            ]);
68      waitbar(i / length(videodata));
69   end
70   close(wait)
71
72 disp('File Processing Finished');
73 tEndOverall = toc(tStartOverall);
74 disp(['Processing Time: ' num2str(tEndOverall) 's']);
75
76 %%% E N D   O F   F I L E
```

C.2   File Processing—Command-Line Commands

Listing C.4 Netem and ffmpeg commands for packet loss processing for video

```
1  // adds 0.5% of random Packet Loss
2  sudo tc qdisc add dev lo root netem loss 0.5%
3
4  sudo tc qdisc show
5
6  // delets the set PL
7  sudo tc qdisc del dev lo root netem loss 0.5%
8
9  // // Streaming:
10 // Receiver(Terminal 1):
11 ffmpeg -y -i udp://127.0.0.1:1234 -c:v copy -crf 25 -g 25
       RECEIVEDFILE.avi
12
13 // Sender (Terminal 2):
```

```
14   ffmpeg -re -i SENDFILE.avi -c:v libx264 -preset slow -pix_fmt
         yuv420p -crf 18 -g 25 -tune zerolatency -f avi udp
         ://127.0.0.1:1234
```